The impact of micro-electronics

The impact of micro-electronics

A tentative appraisal of information technology

J. Rada

International Labour Office   Geneva

ISBN 92-2-102378-8 (limp cover)
ISBN 92-2-102383-4 (hard cover)

First published 1980

Photocomposed in India
Printed in Switzerland

PREFACE

During the past few years further rapid technological advances have been made, particularly in micro-electronics, which are bound to have far-reaching effects in almost every sphere of economic activity and on a large number of occupations. At a time when the employment situation has aroused serious concern in all market economy countries, it seems particularly appropriate that considerable attention should be paid to ascertaining the relationship between these major technological changes and employment. While the rate and extent of adoption of micro-electronics are still debatable, it can safely be assumed that micro-electronics will in any event lead during the 1980s to the emergence of a large variety of new products and to changes designed to save capital, labour, materials and energy.

The effects of these changes will not be limited to the advanced industrialised economies. Developing countries will feel them also, and it can reasonably be supposed that these effects will have considerable influence on the existing international division of labour. Alert to this situation, both the International Labour Conference of 1979 and the European Regional Conference held in the same year urged the ILO to study as soon as possible the employment implications of the recent technological developments. This book is the first of several contributions that the ILO will be making to the discussion of this issue, which is of growing concern to workers, employers and governments in both advanced and less developed countries.

By examining the spread of micro-electronics in a global context, the study breaks new ground inasmuch as the possible consequences for the developing countries have hitherto received little attention. These effects could lead to the erosion of the comparative advantage of those countries in international trade. The trends and tendencies described in this study raise a number of issues of primary importance for the North-South dialogue and create a further challenge in the debate on alternative development strategies.

Francis Blanchard,
Director-General,
International Labour Office.

CONTENTS

Terminology

Definitions of many of the technical electronic terms that occur in the text are to be found in the *Collins Dictionary of the English Language* (London, 1979). In other cases an attempt has been made to clarify the meaning either in the text or in the notes.

Acknowledgements

This study could not have been produced without the purely voluntary assistance of Mrs. Peggy Allemand; the author is especially grateful to her. He also wishes to express his thanks to Mrs. Liz Hopkins, who assisted him in the writing and streamlining of the text, and to Luis Ramallo of the Facultad Latinoamericana de Ciencias Sociales for his valuable advice.

INTRODUCTION

Recent developments in the electronics industry have sparked off a debate in the developed countries on the socio-economic effects of new technologies. The aim of this study is to draw together the information available on industrialised market economy countries and to extend the debate to include effects on the developing countries and on the international division of labour, both of which have been rather neglected hitherto.

The recent developments just referred to are centred on the creation of integrated silicon microcircuits. These have acted as the catalyst in the synthesis of electronics, computing and telecommunications technologies. The raw material which these technologies have been developed to handle is information, and the collective name for this technological trinity is "information technology" or "telematics".

The extraordinary significance of information technology has led some to compare it to the wheel, the steam engine and the mastering of electricity. These impressive claims are based on the technology's role in enabling information to be handled and processed quickly, economically and, most important of all, through a coherent system using a uniform signal, the electronic impulse ("bit"). The information-handling capabilities of the technology have reached a point at which it can perform and extend a wide range of functions that were previously only within the reach of human intelligence.

In this study information is regarded as a resource, in much the same way as energy is a resource. Information is an essential component of mechanical and intellectual activities, and all systems operate through some form of classical information exchange: input-processing-output. It is thus easy to see why information technology is so pervasive; it can ultimately transform all the machinery by which societies produce and service and govern themselves. Coupled with the nature of the technology are its wide availability, exceptional reliability and, above all, the tremendous cost reductions it promises. A short description of the celebrated "chip" will show why.

The transistor, which is the immediate ancestor of today's integrated circuit, pioneered "solid-state technology" by using semiconductor materials. Semiconductor devices can amplify, switch and store electrical signals (bits). The silicon integrated circuit is the most important of the semiconductor

1

devices, containing many interconnected transistors and other components on one "chip" of silicon measuring a square centimetre or less.

The number of transistors on one chip, or the scale of integration, has been central in recent developments. From 10 components per chip in 1960 to 64,000 in 1978, forecasts predict 1 million transistors per chip by 1985.[1] This rapid increase in the scale of integration has been accompanied by the introduction of microprocessors and microcomputers (a microcomputer is a complete computer, including processor, memory and interface, on one chip).

While the chip is at the heart of present developments, the use that can be made of it is conditioned by innovations in other respects, for example with regard to sensors, peripherals, language and software, and communications. At the same time, changes in these fields are becoming dependent on the development of micro-electronics. It is thus not surprising that microcircuits, computers and communications are coming to form a single integrated economic sector – the information industry – and that developments in any one part of it produce a chain reaction in the rest.

The tremendous development and pervasiveness of micro-electronics has transformed electronics into a "convergence" industry, playing a role similar to that of the machine tool industry in the nineteenth century, radiating innovation and permeating other sectors while making electronic skills essential in many other industries.[2] This shift in the locus of innovation has manifold consequences for skill requirements, materials and design. In fact, the electronics industry will be essential to the future.

In the report on the OECD Interfutures Project, established to study "the future development of advanced industrial societies in harmony with that of developing countries", the development of microprocessors has been described as "a decisive qualitative leap forward", and it is stated that ". . . the electronics complex during the next quarter of a century will be the main pole around which the productive structures of the advanced industrial societies will be reorganised."[3]

The nature of the technology and its economic attractions means that it can be applied to a wide range of products and processes. It intervenes in both productive and administrative activities. It can be used to relate weight to price in a shop, control the cycles of a machine tool, guide an aircraft to a safe landing or optimise the mixture of fuel and air in the carburettor of a car. It will significantly alter production processes by increasing automation through micro-electronic controls and the development of robots. It will have a significant effect on the way services are produced, and on the way clerical work is performed, owing to the introduction of word processors, electronic storage systems and office automation in general. Thanks to micro-electronics, automation is penetrating liberal and artistic professions (e.g. medicine and design). It will bring about an exponential increase in scientific and technological research capacity. The technology will encourage the use of computer terminals in the home, the transmission of textual information through the telephone network and satellites, the development of electronic banking and remote control of factories and equipment.

Since there would appear to be no insurmountable technological difficulties, the extent and types of application depend largely on social and political considerations.[4] The current debate is centred on two major socio-political themes: the ownership and flow of information or data and labour displacement effects.

The question of the ownership and flow of data raises problems of international dependence, or interdependence, created by trans-border data flow (the sharing of data or computing resources between countries), which may result in making a country vulnerable to events and decisions in another country, outside its control. In addition, the concentration of data processing capacity in one country results in a loss of economic opportunities, research and development facilities, and control over the data by other countries. It is also difficult to safeguard the privacy of personal, corporate and financial data, both private and public, which can be transferred to or retrieved in countries that may have lower standards of privacy and security than the country where the data originated. The problem becomes more complex when it is remembered that the prime users of trans-border data flow are multinational companies transmitting an intangible good which it is almost impossible to monitor. Because of the increasing influence of those companies in the trade and industry of many different countries, data relevant to the day-to-day functioning of a country can be held outside its borders.[5] Various other developments are improving the efficiency and scope of surveillance techniques, record keeping and other forms of social control.

The debate on the labour displacement effects of the technology is beset with forecasting difficulties. However, the evidence seems to suggest the beginning (or for some observers the continuation) of a shift from a society with unemployment to one in which increased leisure is regarded as normal and the entire potential labour force is no longer needed to produce the necessary goods and services. There is general agreement on the benefits of the technology and the need to apply it quickly in the industrialised market economy countries in order to maintain competitiveness. Disagreement arises over whether or not new job creation will keep pace with labour displacement.

The discussion of the effects of technical change on society and labour is certainly not new. The Roman emperor Vespasian (69–79 AD) is said to have opposed the use of water-power because he felt that it would create unemployment.[6] It could be argued that we are in the midst of another "automation fever", similar to that which followed the commercialisation of the first computers in the 1950s; but there are strong indications that this is not in fact the case today. Technological determinism was responsible for the automation fever which broke out when the first computers arrived and which is reflected to some extent in today's debate. In the present study, however, an effort is made to point to economic, social, and cultural factors that condition the diffusion and application of the technology; although a few are simply mentioned, others are treated in some detail.

A failure to consider all intervening factors in the relationship between technological and scientific change and society might lead to the conclusion that since computerisation did not produce the foreseen alterations in the past it would not produce them in the future. Such a conclusion would follow from an erroneous essential assumption; it supposes that technologies and socio-economic conditions are static. Taken to its logical extreme, such an assumption

3

could lead one to maintain, for instance, that a particular natural resource would not be scarce in the future because it had not been in the past. Let us look at a different situation from history, which supports the more dynamic approach taken in this study. Cumulative changes in steam technology and metallurgy were responsible for the transition from the atmospheric engine of the early eighteenth century to the high-pressure engine at the beginning of the nineteenth century. It was the latter engine that made the economic and widespread use of steam possible and contributed to a profound change in human existence. We are facing a similar situation today, when cumulative changes in electronic, computing and communications technology are leading to a change different in both kind and degree from past innovations.

On the other hand, mechanical extrapolation from technological innovations should not be dismissed out of hand. It has served as the favourite technique of science fiction and could also serve as an indicative tool for long-term technological forecasting. There are abundant examples of the usefulness of this technique. Most striking, in terms of the development of information technology, is the fact that of the 137 devices imagined by George Orwell some 30 years ago in his book *1984*, about 100 are now practical.[7]

There are other theoretical and empirical limitations which make it difficult to put forward firm predictions about such issues as labour displacement. As has been mentioned, information technology is invading practically all fields, so that assessment becomes hazardous. There is also a lack of a proper understanding of the effects of science and technology on society—more specifically, for the purpose of this study, an understanding of the relations between technological change, production, labour effects and the international division of labour. In the numerous works that have been written in this field the tendency has been to treat technology as an exogenous variable or as a residual, dynamic factor of production incorporated in labour or in capital or in both. A comprehensive framework and explanatory variables which would assist in an assessment of particular technological developments is not available. Co-operative research and open debate is needed to fill this gap.

Furthermore, it is extremely difficult to isolate the introduction of micro-electronics in either products or processes from organisational or other changes which generally accompany them. Predictions have been made for some sectors in particular countries, but they should be regarded as indicating little more than orders of magnitude, owing to the methodological problems already mentioned. Therefore we have to rely on subsequent evaluation of experience in selected industries and services, predictions for certain sectors and, perhaps more important at this stage, an understanding of the potential use of the technologies.

These difficulties hinder the formulation of clear recommendations on the innumerable issues involved. This is not to say that immediate action should not be taken but to emphasise the need for debate, research and above all a sufficiently long view of the social and political implications.

In spite of the limitations mentioned, it is possible broadly to sketch likely trends on the basis of evidence accumulated in many different sectors and fields. No technology is independent of developments in other theoretical, scientific and technological fields, nor of socio-political and economic conditions and the prevailing philosophy of science, the proper framework of analysis being the effect of science and technology on society as a whole rather than in a more limited field. In that sense, this study is a tentative appraisal of an extensive subject which needs urgent consideration.

The initial sections of this study are devoted to a discussion of the uniqueness of the new technologies and their technical characteristics, which serve as the framework for assessing their effects. On the basis of current trends and existing data, attention will be focused on the speed of diffusion of the technology, the different lines of products developed and their potential uses. The information available in these respects (mainly concerning the developed countries) will be used together with past evidence in examining the possible effect of information technology on developing nations as well.

Notes

[1] United Kingdom, Cabinet Office, Advisory Council for Applied Research and Development (ACARD): *The applications of semiconductor technology* (London, HMSO, 1978); and *Scientific American* (New York), Sep. 1977; see also *Semiconductor scene meets the microprocessor, A New Scientist* publication (London, 1978).

[2] On this point see Staffan Jacobsson: *Technical change, employment and technological dependence*, University of Lund Research Policy Institute, Discussion paper No. 133 (1979), p. 24.

[3] Organisation for Economic Co-operation and Development (OECD): *Interfutures: Facing the future: Mastering the probable and managing the unpredictable*, report based on the research of an international team . . . under the leadership of . . . Jacques Lesourne (Paris, 1979), pp. 114, 336.

[4] See Iann Barron and Ray Curnow: *The future with microelectronics: Forecasting the effects of information technology* (London, Frances Pinter, 1979).

[5] Simon Nora and Alain Minc: *L'informatisation de la société*, Rapport à M. le Président de la République (Paris, La Documentation Française, 1978), pp. 71-72, and ibid., *Annexes*, Vol. 1: *Nouvelle informatique et nouvelle croissance*, Annex No. 2: Raimundo Beca: "Les banques de données". See also the series of Informatics Studies and Information, Computer, Communication Policy (ICCP) publications of the Organisation for Economic Co-operation and Development, particularly No. 2 in the latter series: *The usage of international data networks in Europe* (Paris, OECD, 1979).

[6] T. K. Derry and Trevor I. Williams: *A short history of technology from the earliest times to A.D. 1900* (Oxford University Press, 1970), p. 252.

[7] David Goodman: "Countdown to 1984: Big Brother may be right on schedule", in *The Futurist* (Washington, World Future Society), Dec. 1978, pp. 345-355.

THE NATURE OF THE NEW INFORMATION TECHNOLOGY

1

Micro-electronics is now the technical basis of information technology. In handling and processing information, that technology enlarges in a quantitative sense the capacity of human intelligence and acts to some extent as a substitute for it. These are the abstract characteristics of the technology, and they constitute its nature, its essential and fundamental concern.

The chief machine in which the technology is embodied is the computer. The "hardware" of a computer consists of input and output devices, a central processing unit and a memory. The computer accepts information from the environment through input devices. It combines the data according to the rules of the programme stored in its memory and returns the results to the environment through output devices. It cannot operate without a system of instructions, or "software".

The parallel between human beings and computers has often been drawn, human sensory systems being the input unit, the brain the central processing and memory unit and speech or action the output system. The computer, like the human brain, is a universal information-processing machine in that it can handle any information compatible with its programme. New types of information can be accommodated by re-programming the computer instead of inventing a new machine. The nature of the intelligence functions performed by computers has been debated since their early days. The information processing capability of the machines allows them to set goals and sub-goals, make plans, consider hypotheses, recognise analogies and certain patterns; they are able to link different facts and carry out other intellectual activities, while at the same time reacting to changes in their environment and thus adapting to changing circumstances. Furthermore, the programming of a computer is theoretically comparable to the way in which a large part of human behaviour is causally conditioned by genes, parents, teachers, the physical environment and so on.

It is clear that machines cannot "think" in a human sense. However, computers are able to carry out the functions described in the preceding paragraph. The debate on whether these functions can be said to be "intelligent" or constitute "thinking" is a semantic minefield.[1] Probably because they originated in calculating machines, computers have so far been used almost exclusively for calculations. There is no reason, however, why they should not handle other forms of information and play a larger role in connection with other human and mechanical functions. The key to such an extension of their use is the development and improvement of appropriate input systems to transform different signals into digital language, or strings of one and zeros, so they can be manipulated by the computer.

Information is an inexhaustible resource. Unlike most commodities, its amount is not reduced by repeated use, and its value can be increased by its circulation. The universal availability of information would open the whole field of knowledge to non-specialists. Information is also a vital link between the most diverse mechanical and intellectual activities. A report issued by the OECD in 1971 contains a list of areas in which information is a resource: ". . . computing, publishing, newspaper, broadcasting, library, telephone and postal services . . . , together with large slices of teaching, of government, or industrial and commercial operations, and of many professional activities."[2]

No consistent theory has yet been developed to enable us to grasp the multidimensional aspects of information. However, an understanding of the computer as a universal information machine and of the characteristics of information itself gives us clues to the nature of the current technological changes and how they differ fundamentally from past technological evolution. The slow accumulation of changes towards a surplus in agriculture, the development of trade and the influence of new ideas and concepts of nature created, over centuries, the basis for the Industrial Revolution. That revolution in turn, within a very short time, has provided the foundation for what has been termed the "information revolution". It can be said that, from a technological point of view, the Industrial Revolution was characterised by the extension of human muscle power and dexterity through the mastery of universal and convenient forms of energy and mechanical devices. This is not to diminish the importance of innovations in areas such as materials, chemicals and metallurgy, but to emphasise the main features. David Landes defines the technological side of the Industrial Revolution as ". . . substituting machines for human skill and inanimate power for human and animal force . . .", which "brings about a shift from handicraft to manufacture and, so doing, gives birth to a modern economy".[3] Nevertheless, during the industrial age, human intelligence functions continued to be performed exclusively in the human brain.

Towards the end of the nineteenth century, Taylorism and scientific management attempted to withdraw these functions from the shop floor and place them in a planning department of management, thus polarising knowledge and skill within production. This development was in conformity with a more profound tendency of capital to become as independent as possible of the human factors that condition its reproduction. Automation, the star of the micro-electronic universe, can be seen as the tendency to withdraw as many human elements as possible from production (manufacturing and administration), owing to the unreliability of what an author in the United States has called "human operating units".[4] This "unreliability" includes the relative slowness of human beings in the performance of certain operations, the cost of their labour and their political propensities. The introduction of micro-electronic technology, then, is the continuation of an earlier process but on a tremendous scale and at a greatly accelerated pace. The continuity is contained in the progressive fragmentation of human labour. At this stage machines are being substituted for human intelligence, much as they were substituted for human muscle power in an earlier epoch. While man's first technological breakthrough related to ensuring survival, and the second to multiplying muscle power, finding substitutes for it and improving manual dexterity, the third breakthrough, which is occurring today, concerns the extension of the possibilities of applying human intelligence by the use of a more efficient substitute for some of the functions of the brain. From a different angle, it has been argued that in terms of biological evolution, the first important threshold in the development of the energy-matter-information relationship occurred in the carboniferous period, when an organism emerged which, for the first time in history, had more information in its brains than in its genes.[5] The second jump occurred about 5,000 years ago with the invention of different symbols and characters which allowed information to be stored elsewhere than in human memory. The introduction of printing marked a quantitative development of this trend.

Microprocessors may be considered as the third stage of evolution. For the first time in human history, not only extrasomatic storage of information but also extrasomatic storage of intelligence became possible. This is definitely a qualitative change. The acceleration in the sequence of these three stages—the first of which did not involve . . . *homo sapiens*—is breathtaking: hundreds of millions of years, thousands of years, and finally, decades.[6]

From yet another point of view, Daniel Bell maintains that human societies have seen four distinct revolutions in the character of social interchange: speech, writing, printing and now telecommunications. The newer changes constitute the basis for the development of the "information society".[7]

While the foregoing descriptions have placed the current information revolution in the framework of human evolution, what are the direct technical antecedents and what sets of conditions generated their invention and use? The demand for information handling and processing grew apace with the Industrial Revolution. The burgeoning of trade generated improved communications through the establishment of railway and shipping networks, and efficient telegraph and telephone systems. The increasing complexity of society, requiring systematic correspondence and record-keeping, led to the introduction of the typewriter, which was in extensive use by the 1880s. Equally important was the need for a machine capable of processing numerical information. In 1822 Charles Babbage introduced the difference engine, which was used for the computation of tables. In 1833 he began work on the design of a steam-powered "analytical engine". Not surprisingly the latter design, which was based on theories years ahead of engineering technology, was never successfully developed. Punched cards, used by Jacquard in the late eighteenth century to control patterns woven by the automatic loom, were to have been employed to direct the activities of the new machine, which would have been able to retain and display on demand any of its thousand 50-digit numbers (thus incorporating a memory for the first time). It would have been possible for the machine's operations to be adjusted automatically according to the results of its own calculations.

In 1890 Herman Hollerith put punched cards to use in his electric accounting machine, which was used in the United States census of the same year. This was not a computer but a machine which sorted and collated cards according to the position of the holes punched in them. In 1946 International Business Machines (founded in 1911 under the name of Computing-Tabulating-Recording Company) announced the first electric calculator, which was capable of performing arithmetical operations on data contained on punched cards. From 1943 to 1950 several data processing machines were developed at Manchester and Harvard universities, and the first commercially available computers were put on the market in 1950.[8] As in the case of the steam engine, initial development of the computer was slow. Because the first devices were large, unreliable and expensive to purchase and operate, they were available only to large enterprises. The electromechanical relay computer or the vacuum tube computer can be compared to early versions of the atmospheric engine; the transistor computer to Watt's steam engine; and the micro-electronic computer to the high pressure engine of 1800 which made the use of steam more widely available. Micro-electronics has made it possible to put computers to every use.

One of the results of the Industrial Revolution was the marriage of science and technology which, in turn, has served as the intellectual basis of the information revolution. Commenting on the information revolution, Daniel Bell states:

The first "modern" industry is chemistry, in that the scientist must have the knowledge of the theoretical properties of the macromolecules that he is manipulating in order to know where he is going. What is true of all the science-based industries of the last half of the twentieth century, and the products that come from them—electronic equipment, polymers, computers, lasers, holograms—is that they derive from work in theoretical science, and it is theory that focuses the direction of future research and the development of products.[9]

At the same time, and as with the steam engine, computerisation lays down a particular challenge to other areas of science and technology, and it points to the need for further developments in such fields as sensory devices, electron, x-ray and laser technology (opto-electronics),[10] materials, pneumatics, hydraulics and the study of electrical and magnetic properties. It has also shown the need for a more systematic theoretical consideration of information.

The technical and intellectual antecedents of the information revolution are clear, but what is the impetus behind the current changes? It is important to realise that in most countries it was only in the early 1950s that national accounts became available on a regular basis. Only after the Second World War were factual data such as trade and production statistics and the results of market surveys and opinion polls systematically collected on a large scale. This was in recognition of the need for more precise forecasting and adjustment of economic and social trends, and the number of decisions based on numerical information increased at an accelerated rate each year. At the same time, decision-making became global in scope. National policies have world-wide ramifications, and economic, social, cultural, environmental, technical and scientific issues cross national boundaries. New theoretical and practical difficulties arise because the resource used in these types of international exchange is information, which is intangible and difficult to conceptualise and evaluate. For example, the concept of money supply has become increasingly abstract as it has expanded from coinage to cover paper money and now the electronic transfer of funds and the availability of funds through such devices as credit cards.

The consequences of the need for information and its global use have inevitably included startling increases in its quantity. In the technical and scientific field, a 1973 study of the Organisation for Economic Co-operation and Development (OECD) forecast an annual 12.5 per cent growth in the number of documents; accumulation at this rate will mean a general stock of 120-150 million documents by 1985-87.[11] Indeed in the space of three centuries (1660-1960) the volume of scientific documentation has increased by a factor of 1 million, and this type of material tends to be used only by a small, elite market.[12] The OECD study already referred to stated:

> Before we are overwhelmed by this rising tide, we will be able, for some time yet, to evade the issue, look for means of escape and adopt stopgap solutions. But after 15 or 20 years the choice must inevitably be between automation and suffocation. In these circumstances there can be no doubt as to the outcome.[13]

A further impetus to the rapid development of micro-electronics has been the need to increase productivity in the office now that economic growth can no longer be achieved through the increase of industrial and agricultural productivity alone.

Thus the universal need to process and use information, the quantity being generated and its rising importance as a factor in economic growth provide the conditions to which micro-electronic technology responds.

An illustration of the shift into information-related economic activity is given by some studies undertaken with regard to the United States. According to one report, information occupations accounted for over 40 per cent of the total labour force by 1970.[14] Another source (1979) maintains that the information industry accounted for one-quarter of the gross national product in 1955, and that in 1979 it would account for more than half.[15] Although statements such as these rely on debatable definitions which need to be further refined, they point crudely to the basic contemporary trend.

The shift is also reflected in the infrastructure for the transmission of data. In Europe alone, there is expected to be a twelvefold increase in the volume of data traffic between 1975 and 1985.[16]

Dependence on information is such that the withdrawal of information machines (computers) would entail virtual paralysis of the developed economies and of their capacity to govern and defend themselves. Manufacturers of data-processing equipment maintain that they will still be producing and selling in profitable quantities until at least 1985, even without taking into account the decreasing cost of hardware, because the accelerated increase in demand for the recording, storage, analysis, processing, retrieval and communication of information is and will be so enormous.[17]

Notes

   [1] For details of the above-mentioned aspects, see: A. M. Turning: "Computer machinery and intelligence", in *Mind* 39, 1950, pp. 433-460; Marvin L. Minsky: "Artificial intelligence", in

Scientific American (New York), Sep. 1966, pp. 247-260; F. H. George and J.D. Humphries (eds.): *The robots are coming: The implications of artificial intelligence developments* (Manchester, National Computer Centre, 1974); Barron and Curnow: *The future with microelectronics*, op. cit.

[2] Organisation for Economic Co-operation and Development, Committee for Science Policy: *Report of the ad hoc group on information computers and communications* (1971).

[3] David S. Landes: *The unbound Prometheus: Technological change and industrial development in Western Europe from 1750 to the present* (Cambridge University Press, 1969), p. 1.

[4] R. Boguslaw: *The new utopians: A study of system design and social change* (Englewood Cliffs, New Jersey, Prentice-Hall, 1965), as quoted in Mike Cooley: "Contradictions of science and technology in the productive process", in Hilary and Steven Rose (eds.): *The political economy of science: Ideology of/in the natural sciences* (London, Macmillan, 1976), p. 88.

[5] Carl Sagan: *The dragons of Eden* (New York, Random House, 1977), p. 47, as cited in Bruno Fritsch: "Some socioeconomic implications of microprocessors in an evolutionary perspective", in International Social Science Council, European Coordination Centre for Research and Documentation in Social Sciences (Vienna): *Micro-electronics and macro-socio-economic issues: Employment consequences,* Theme b. 1, Microelectronics conference, Zandvoort . . ., 1979 (Documentation II, WD 2 (Rev.); mimeographed), p. 43.

[6] Fritsch, loc. cit.

[7] Daniel Bell: "Communications technology–for better or for worse", in *Harvard Business Review* (Boston), May-June 1979, p. 20

[8] Keith S. Reid-Green: "A short history of computing", in *Byte* (Peterborough, New Hampshire), Vol. 3, No. 7, pp. 84-94.

[9] Bell, op. cit., p. 22.

[10] "Opto-electronics" refers to the interaction between optics and the properties of light, on the one hand, and electronics on the other. It is a generic term covering a wide area which includes laser technology, optical fibres, image-based sensors and related devices.

[11] Georges Anderla: *Information in 1985*, A forecasting study of information needs and resources (Paris, OECD, 1973), p. 120.

[12] D. J. de Solla Price: *Little science – big science* (New York, 1963), p. 9.

[13] Anderla, op. cit., p. 89.

[14] Edwin B. Parker: "Background report", Session A: "Social implications of computer/telecommunications systems", in *Conference on computer/telecommunications policy*, Proceedings of the OECD conference, . . . (Paris, 1976), p. 93. Parker's definition of the "information sector" (ibid., Appendix: "The labour force aggregation scheme", pp. 118-129) was worked out with the assistance of Marc Porrat, author of *The information economy* (Institute for Communication Research, Stanford University, unpublished paper, 1974). According to that definition the sector consists—

(a) mainly of some of the occupations that were classified by the United States Bureau of Labor Statistics as being in the service sector (engineers, scientists, teachers, managers and administrators, members of the professions, financial and marketing staff and secretaries, but excluding occupations in the health sector, technicians and salesmen); and

(b) of a very small proportion of members of the labour force classified by the Bureau as being in the industry sector (mostly telephone and broadcasting installers and repairmen and workers in printing and kindred trades).

[15] International Data Corporation: "Productivity and information management", advertisement in *Fortune* (Chicago), 12 Mar. 1979, p. [28].

[16] H. P. Gassman: "New international policy implications of the rapid growth of transborder data flows", in OECD: *Transborder data flows and the protection of privacy*, Proceedings of a symposium held in Vienna, . . . 1977, Information Computer Communications Policy Series, No. 1 (Paris, 1979), p. 55.

[17] T. A. Dolotta et al.: *Data processing in 1980-1985: A study of potential limitations to progress* (New York, John Wiley, 1976), p. 33.

CHARACTERISTICS AND DIRECT COST

2

The advances in micro-electronics have been based on the increased complexity of silicon integrated circuits. Microcircuits integrate many of the interconnections previously required in electronic components, making them more reliable and reducing maintenance and materials costs. They also operate on a low voltage and in a normal (i.e. not air-conditioned) environment. There are several types of integrated circuits: they range from those which are mass-produced and used in such articles as pocket calculators, television games and watches to high-performance, custom-made products. Accordingly costs, prices and manufacturing techniques vary enormously. However, it should be noted that on the whole, although the cost per chip has remained constant or even increased, the number of components per chip has grown exponentially, so that there has been a reduction in cost per function. Costs per bit have been reduced about a thousand times since 1965, and further reductions of 25 to 30 per cent a year are expected in the future. This reduction will depend on the density obtainable and on improvements in manufacturing techniques. Pricing policy and the cost of research and development will also determine the final price of the circuits. Today's microcircuit technology enables an individual integrated circuit on a silicon chip about half a centimetre square or smaller to embrace more electronic elements than the most complex piece of electronic equipment built in 1950. Today's silicon chip holds more computing than the first commercially available computer: it is 20 times faster, has a larger memory, is thousands of times more reliable, consumes the power of a light bulb instead of a locomotive, occupies 1/30,000th of the volume and costs 1/10,000th of the price.[1] The density of the circuits on a chip is expected to be increased by the use of electron beam and x-ray patterning, which will supersede optical lithography. The new systems could conceivably reduce the linear dimensions of a transistor by a factor of 50, and the area occupied by a factor of more than 1,000. Predictions for very large scale integration (1 million elements per chip) are based on these new manufacturing techniques.[2] Some of these new techniques will include the use of different semiconductor materials such as gallium and arsenide, the commercial exploitation of the Josephson superconductor[3] and the development of complementary technologies such as the use of magnetic bubbles[4] and coupled charged devices. Improvements in manufacturing are also expected through automation and better design. These may increase yields per wafer from about 25 per cent in 1976 to 90 per cent in the early 1980s.[5] Cost reductions in packaging and assembly are also taking place thanks to highly automated techniques.

The economy of micro-electronics in terms of performance for a given price is the single most important factor that explains its pervasiveness. Chips are being incorporated in consumer goods, machine tools and electrical and electronic equipment. The control element in the information technology revolution is thus becoming cheaper while the number of functions it can carry out is increasing. Increases in performance in relation to price are shown by IBM's disclosure that in 1977 it had orders for computers totalling $11,000 million, which represented 20 per cent of the value of all IBM computers installed in the world. Moreover, the orders represented four times the processing power of all the computers that IBM had ever delivered.[6] Table 1 compares price and performance of a 1955 computer and a 1978 calculator respectively.

Table 1. Comparison of the characteristics of a 1955 computer and a 1978 calculator

Characteristics	A 1955 computer (IBM 650)	A 1978 calculator (TI-59)
Components	2 000 vacuum tubes	166 500 transistor equivalents
Power (KVA)	17.7	0.00018
Volume (cu. ft.)	270	0.017
Weight (lbs)	5 650	0.67
Air conditioning (tons)	5 to 10	None
Memory capacity (bits):		
Primary	3 000	7 680
Secondary	100 000	40 000
Execution time (milliseconds):		
Add	0.75	0.070
Multiply	20.0	4.0
Price (current dollars)	200 000	300

Source: Texas Instruments Inc., shareholders' meeting report, 1978.

The purely physical characteristics of microcircuits indicate the economic attractions of the new technology to the developed world, where savings in energy and space are becoming more and more important. The functional characteristics of microcircuits offer further economic advantages. The high versatility and flexibility of the basic components allow microprocessors or microcomputers to be programmed to perform different functions. These micro-electronic elements can be used to intervene in both production and organisation. Micro-electronics can be used for such purposes as the allocation and use of resources, stock control, warehousing, distribution and production. Micro-electronics allow automation of monitoring and control. The time and cost of gathering and processing information are reduced, and efficiency is improved, by more rapid reporting and decision-making, closer monitoring of markets and increased flexibility of services. In terms of capital, there is a saving per unit of output although an increase per worker employed.[7]

The current and future reductions in the cost of microcircuits and computers as a result of higher densities, new materials and new manufacturing techniques do not imply similarly dramatic reductions in the total cost of the equipment. Hardware represents only a part of the general cost of computers and related equipment; other elements are software, or programmes, and peripheral equipment such as terminals, printers and modems. More often than not, installation of equipment also has to be accompanied by changes in organisation, data-gathering procedures and shop floor design.

LANGUAGE AND SOFTWARE

Computers can operate only if they have complete instructions on how and when to react to inputs. The instructions are provided by various types of programmes, collectively called "software". Software may account for up to 80 per cent of the total cost of equipment, and is therefore expensive by comparison with hardware. Programmes can be written in different computer languages; the complexity of a language is an essential feature of a programme and its performance and determines its application. There are three main lines of development in this respect. One trend is the incorporation of software into hardware through the manufacture of a pre-set computer with built-in programmes; this product is termed "firmware". Firmware is important in the purpose-built market sector. The second main trend is towards using programming languages that are easy to master though offering increased complexity. An example of these languages, intended for mass use, is BASIC (an acronym for "beginner's all-purpose symbolic instruction code"). BASIC expresses programmes in a conversational manner; input is sequential and fed directly into the computer, which carries out elementary checking and validation of each step before the next one is introduced. In this way, BASIC incorporates extensive diagnostic and correcting facilities. As a result of the availability of this type of language, computers have become consumer goods. The third main line of development is an increase in the productivity of programmers through automation and the mass production of programmes.

PERIPHERALS

The peripheral equipment to support, convert or translate data and to link up different kinds of computers has also been revolutionised. Two broad categories of peripherals can usefully be distinguished: input-output devices and data-storage devices.

The recording of information and its translation into electronic signals remains a difficult and relatively expensive operation, slowing down the pace of present changes. Human output is primarily verbal, while computer input is mainly numerical through manually operated terminal keyboards (punched cards are used only in older machines or for special applications). Closing this gap remains one of the challenges to the industry, but it is generally agreed that direct verbal input of the complexity required is far in the future.[8] The most common type of terminals are composed of a keyboard and a cathode ray tube television screen or video display unit. Today there are an enormous variety of terminals, from audio-input terminals the size of a pocket calculator, through portable terminals the size of an executive briefcase, to "intelligent" terminals with analytical and processing capabilities of their own. The terminals most familiar to the general public are those used by travel agencies and airports for real-time computing. The use of a telephone with a television set heralds the arrival of the home computer terminal, as under the Prestel videotex[9] system which is already commercially available. The push-button telephone, and the more versatile electronic telephone, are also potential computer terminals.[10]

A number of other input-output devices are used to feed, retrieve and produce information. One of these is the printer. Printers have changed considerably over the last few years, and a modern printer can produce 6,000 lines or more per minute. A whole new industry has grown up in this sector of the data processing world, with special equipment for separating and collating the sheets of printer output. Although further progress is expected in this field, printers remain essentially electronically controlled mechanical devices. Other printer systems will be important in the future, particularly in office automation. Computer output microfilm presents digital data on microfilm, printing at speeds of up to 70,000 characters per second. For a number of applications, this is much more convenient than paper printers; a paper copy is easily obtainable from the microfilm. Computer input microfilm, which extracts digital data from microfilm, remains an exotic technology but should become cheaper and more reliable in the future. There are other input systems that use optical and laser beam scanners to translate textual information into digital language. Facsimiles are becoming important as input or output devices and for communication between offices, paving the way for the more versatile electronic postal system. Computers are also being used to deal with alphanumerical information fed to them through sensory devices. There is still a need for the development of such devices for industry, particularly in the chemical industry and where robots are used.[11]

The storage of information is becoming increasingly important and will develop quickly in the future. The development of data banks and archive information systems, for example, depends largely on this sector of the information industry. Information can be stored in electronic memory devices such as micromemories and magnetic bubbles, or on tapes, drums, diskettes[12] and so on. The systems available differ in their characteristics and applications, the latter being determined by price, storage capacity, access speed and durability. The storage cost per bit is at present about $0.01 (high-speed random access memory) and is expected to fall to 10^{-8} cents by 1985 with the introduction of the video disc (a double-sided pre-grooved plastic disc on which information is engraved and retrieved through a laser beam head).[13]

Storage systems combined with input-output devices create an integrated information system with the computer at the centre. The now efficient communications links between these devices are an indispensable contribution to the information revolution: the development of this section of the industry has been as important as that of the chip itself; without it, communication between micro-electronic devices and human beings would be inadequate. In order to take full advantage of the information revolution, further development of terminals, which are most important input and output devices, will be crucial. In the long run the widespread use of visual terminals, and the availability of an extensive communications network interconnecting them, could supplant paper as the main medium for information presentation and eliminate the need for many of the peripherals in use today. The cost of peripherals has in any event decreased significantly. As they become more "intelligent", price in relation to performance will be further reduced.

COMMUNICATIONS

It is clear from previous sections that communications are central to the full development of the information revolution. The telecommunications industry is quickly gaining in importance as systems and nations organise themselves around communication networks. The ability of computers and peripherals to communicate among themselves and rapidly transmit information makes a major contribution to the achievement of economy and speed. Current telecommunications rely mainly on the telephone and on telephone lines. These transmit computer output, which is digital in nature, in strings of ones or zeros represented by strings of pulses or no-pulses. Thus a standard letter of 20,000 bits, for example, can be transmitted in 10 seconds. The same is true for any kind of data. In this field too, formerly clear distinctions between telecommunications and data processing are becoming blurred. The telephone wire is a relatively slow medium of data or word communication by comparison with the techniques which will be extensively used in the 1980s. Transmission using package switching, whereby data blocks are transmitted in bulk from station to station (using a method similar to the one used by postal services) and then distributed through networks, will considerably decrease the cost of transmission.

Future advances in communications will depend on the modernisation of telephone networks and a reduction in the price of communication channels and equipment; needless to say these requirements are being attended to. The commercialisation of the videophone is gathering momentum and the cost of launching communications satellites will soon be reduced considerably because of the space shuttle of the United States National Aeronautics and Space Administration; roof-top satellite-to-earth stations are already being used in Canada, India, Japan and the United States.[14] Reduction in cost on the whole telecommunications front can be expected in the future. The use of satellites provides a channel cost falling at the rate of 40 per cent a year; it consists of a distance-insensitive cost coupled with decreasing cost for earth stations.[15] More rapid data transmission (see the figures overleaf) will render the system even more economic and flexible.

Data transmission systems	Bits per second
Telephone lines	9 600
American Telephone and Telegraph data-only network	56 000
Xerox (proposed)	300 000
Satellite Business Systems (proposed)	6 300 000

These systems of varying degrees of sophistication will offer computer and telephone links, document distribution and video conferences with live and still images.[16] The growth of this sector of the information industry is expected to be spectacular and is creating demands for new devices and materials. The most important development in this respect is that of fibre optics technology which carries information by using a laser beam in a tube of self-reflecting glass a few millimetres in diameter. A cable 8 mm in diameter (several fibres) can simultaneously transmit 30,000 telephone calls which would require a diameter of 20 centimetres of conventional copper cable. Fibre optics is more reliable, more powerful and cheaper than copper. In the future this will be the standard cable.[17] This technique is still being used on an experimental basis only, but it is expected that by the mid-1980s it will have penetrated telephone communications, computer interconnections and industrial process control systems, and will be used for normal and co-axial cables. At present, its use is limited because of technical problems which include the difficulty of de-coding the message and the need for repeaters and for an electricity supply to the ringing mechanism of the telephone; however, it is expected that these problems will be solved.

Notes

[1] R. N. Noyce: "Microelectronics", in *Scientific American*, Sep. 1977, p. 65.

[2] In the last 25 years the cost of computations, including both logic and memory, has been falling at an annual compound rate of 25 per cent. With micro-electronics the rate has been accelerated. Many studies explore this point; see for details, Lewis M. Branscomb: "Computer technology and the evolution of world communications", in International Telecommunication Union (ITU): *Speakers' papers*, 3rd World Telecommunication Forum, Geneva, 1979, Part 1: *Telecommunication perspectives and economic implications*, Executive symposium, p. I. 4.2; Noyce, op. cit.; and The Economist Intelligence Unit: *Chips in the 1980's: The application of microelectronic technology in products for consumer and business markets*, Special Report No. 67 (London, 1979). On manufacturing techniques, see W. Holton: "The large scale ingration of microelectronic circuits", and W. G. Oldham: "The fabrication of microelectronic circuits", in *Scientific American*, Sep. 1977. For details on the costs of different possibilities see "Which technology will produce tomorrow's chips?", in *The Economist* (London), 25 Nov. 1978.

[3] At very low temperatures near absolute zero ($-273°C$), some metals lose all resistance to electric current and become superconductors. Because electrons behave like waves as well as particles, they can, through the so-called "Josephson effect", establish a link between two superconductors by penetrating or "tunnelling" through a barrier such as a thin layer of insulation that would be expected to stop them according to the laws of classical physics. Josephson junctions can be switched very rapidly into and out of the superconducting state. With this system power dissipation is very low and switching speed could increase 100 times beyond that achievable with current transistor devices. Difficulties of fabrication and design remain to be overcome before the Josephson superconductor can be considered practical, but the potential exists and could improve computer performance considerably (J. S. Birnbaum: "The future computer", in *McGraw-Hill*

1977 Yearbook of Science and Technology (New York, 1977), pp. 41-42; and *McGraw-Hill Encyclopaedia of Science and Technology* (4th edition, New York, 1977), Vol. 13, pp. 304-305).

[4] A magnetic bubble is an area with magnetisation opposite to that of its surroundings. A magnetic bubble chip usually consists of a non-magnetic substrate which supports a thin film on which the bubbles are stored, a spacer layer, and on top a pattern of T-bars used to move the bubbles. When such a device is subjected to changing magnetic fields the bubbles move from one position to the other. This magnetic shift acts as a register: the bubbles can be read or written (positioned) by special sensing devices at different points of their path. Magnetic bubbles are easier to produce than semiconductors and have about ten times more density, with low power requirements. Memories with densities of 16,000 to 100,000 bits are commercially available. Because bubbles retain their information when the power is off, they may some day supersede other forms of information storage (Birnbaum, loc. cit.).

[5] J. Martin and J. H. Collins: *Distributed processing and data base networks in the 1980's*, Seminar, Savant Institute, Munich, Feb. 1979.

[6] International Data Corporation advertisement in *Fortune*, 5 June 1978, p. 24.

[7] In the Federal Republic of Germany the potential capital-output ratio in the office and data processing sector fell by 16.3 per cent between 1970 and 1977, as compared with a rise of 7.1 per cent in total manufacturing and mining. However, there is an increase in capital intensity for industry and the economy as a whole, particularly with the introduction of additional equipment in offices. For details, see Günter Friedrichs: *A new dimension of technical change and automation*, Microelectronics conference, Zandvoort . . ., 1979, document WD 7 (Vienna, International Social Science Council, European Coordination Centre for Research and Documentation in Social Sciences; mimeographed), p. 5.

[8] The forecasts in this point differ, however; IBM believes that within the decade of the 1980s, speech processing will have advanced in power and cost "to the point where speech messages can be converted to text form and processed directly" (Branscomb, op. cit., p. I.4.3).

[9] Generic name used to describe systems for transmitting information stored in a computerised data base across public telephone lines and displaying it on a television screen at the command of the user. The Prestel service provided by the Post Office in the United Kingdom is known in that country as a "viewdata" system (*Financial Times* (Frankfurt and London), 24 Mar. 1980, Section III: "Viewdata").

[10] The widespread use of terminals will change or eliminate many familiar activities. Long-distance monitoring of factories will become possible, and some traditional office work will become redundant; terminals connected to data banks through the telephone network will revolutionise the printing industry by allowing people to select their reading matter at the push of a button; the postal system, shipping, transport and other current patterns of urban communication will all be transformed, as well as commercial transactions and banking through electronic transfer of funds. In the long run, the use of terminals will significantly reduce the use of paper since letters, memos and invoices will be displayed and stored without hard copy, although hard copy could be produced by using printers and facsimiles. At present there is every indication that the availability of cheap, mass-produced terminals will be essential to these developments. The decreasing price of hardware and the simplification of procedures will make terminals cheaper and more powerful. The main impediment, in terms of cost, is the video display, but progress is envisaged along the lines of solid-state technology, similar to that used in digital displays. The health risks posed by visual displays, which include possible eye fatigue and the effects of radiation from defective terminals, will become more acute as terminals become a working tool for a larger number of people.

[11] See, for instance, "Sensors are delaying the microprocessor revolution", in *The Economist*, 9 Dec. 1978, p. 95.

[12] Or "floppy discs". Small, flexible plastic discs coated with a magnetic material on which data for a computer can be stored (*6,000 words*, A supplement to *Webster's Third New International Dictionary* (Springfield, Massachusetts, Merriam, 1976)).

[13] Martin and Collins, op. cit.; see also note 2.

[14] This development will further encourage the integration of many industries into a single economic sector, namely the information industry. IBM (Satellite Business System), Xerox, American Telephone and Telegraph, Western Union, Fairchild Industries, postal authorities and many other private companies in the United States, Canada, Western Europe and Japan are investing heavily in satellites and related communications equipment. Multinationals such as Exxon have created special subsidiaries for office automation, and it is likely that the Exxon Information System will branch out into the communications and satellite field.

[15] Branscomb, op. cit., p. I.4.3; see also Harold A. Rosen: *Space telecommunications*, paper presented to the International Telecommunication Union, CCIR Commemorative Session, Geneva, 23 September 1979. In relation to the various possibilities and financing of satellites, see Albert

D. Wheelon: "The economics of telecommunications in the century of the satellite", in ITU: *Telecommunication perspectives and economic implications*, op. cit., pp. III.5.1-III.5.5.

[16] *The Economist*, 21 Apr. 1979, p. 114; for video conferences a screen in each branch office would allow widely disseminated staff to talk to each other on closed circuit television.

[17] "Optical system for data communications placed in service in Hawaii", in *ITU Telecommunication Journal*, Oct. 1978, p. 553; *Computerworld*, 30 Oct. 1978, p. 35. For details on the economy and applications of fibre optics, see K. G. Corfield: "Optical fibres in communications", A review of the benefits, in ITU, *Telecommunication perspectives and economic implications*, op. cit., pp. III.6.1-III.6.5.

A WORLD INDUSTRY IN A WORLD ECONOMY

3

A characteristic of information technology which distinguishes it from past innovations is that only in a world market can it be fully and economically used. This is partly because of its nature, as indicated in Chapter 1, but also because of a series of commercial forces which will be examined in this chapter. Naturally, the expansion and behaviour of the electronics market condition manufacturing practices and policy, and these in turn will condition the effect of the technology.

A 1978 study[1] expected world-wide sales of electronic products to be close to $140,000 million in 1979, a 13 per cent increase over 1978, and to reach $200,000 million in 1982. Expansion at similar rates is expected to continue into the foreseeable future as more products and equipment become electronic.

Demand for these products, and their manufacture, is mainly located in the developed countries. This has been shown in recent studies which estimate the current market at about $120,000 million in Western Europe, Japan and the United States, and between $10,000 and $15,000 million in the rest of the world.[2]

The rapid growth of the electronics market is partly due to the fiercely competitive nature of the industry. This is not to say that monopolistic practices do not exist, but to emphasise the fact that this is a technology-based industry, where innovations by one company can win it a substantial share of the market to the detriment of its competitors.[3]

This keen competition has prompted constant reductions in manufacturing costs, especially in the component sector of the industry. Practically all companies are exporters and also have manufacturing facilities in several countries. Components are well suited to this type of arrangement because they have a high value in relation to volume and weight, so that transport costs are low. In addition, tariffs on components are low because they are intermediate products. In the electronics industry there are phases of the production process which are particularly labour-intensive and for which factories abroad have been extensively used. Whereas the first two stages of chip manufacture (the making of masks and wafers) are carried out in the developed countries, the chips are exported to developing countries for central component testing and assembly (wiring and encapsulation). The chips are then re-imported.

A study by the United Nations Institute for Training and Research

Table 2. World consumption and production of integrated circuits, 1976-80
(Percentages)

Countries	1976		1980	
	Consumption	Production	Consumption	Production
United States	54	71	42	64
Japan	22	21	26	25
Western Europe	20	8	26	10
Others	4	–	6	1
All	100	100	100	100

Source: Jean-Marie Fabre and Thierry Moulonguet: "L'industrie informatique: Développement, politiques et situations dans divers pays", Annex No. 7 to Simon Nora and Alain Minc: *L'informatisation de la société*, Rapport à M. le Président de la République (Paris, La Documentation Française, 1978), *Annexes*, Vol. II: *Industrie et services informatiques*, p. 77.

(UNITAR) showed that, in 1971, all major United States manufacturers were also operating overseas, forcing firms incorporated in other advanced countries to follow a similar pattern. The policy of manufacturing abroad has been partly caused and reinforced by sections 806.30 and 807.00 of the United States tarrif schedules which compute import duties on the value added abroad only.[4] The composition of trade in electronic components shows the extent of the use of factories abroad. Imports of the group of products including integrated circuits (thermionic valves and tubes, transistors, etc., item 729.3 of the Standard International Trade Classification) to the industrialised market economy countries quadrupled in c.i.f. value at current prices between 1972 ($346 million) and 1976 ($1,357 million). Six countries—five of them in the Far East—accounted for more than 95 per cent of developing countries' exports. For finished electronic products the figures are equally impressive, and again the same countries account for a high proportion of total exports from developing countries. The imports of industrialised market economy countries from developing countries have increased more rapidly than their exports. However, the trade balance in value against developing countries doubled between 1972 and 1976.[5]

The heart of current developments is the integrated circuit, because of the technical and functional characteristics described in Chapter 2. Integated circuits already represented 52 per cent of all semiconductors in 1976.[6] In the field of integrated circuits the supremacy of the United States is overwhelming: currently the United States accounts for 60 to 70 per cent of world production. It is a highly concentrated industry, with five companies handling 80 per cent of the above-mentioned total: Texas Instruments, Fairchild, National Semi-conductors, Motorola and Intel. Other electronic giants, such as IBM, produce their own chips. World production and consumption of integrated circuits is distributed as shown in table 2.

The market for integrated circuits will continue to expand, but with a much heavier concentration on processing and memory devices. At the same time, the manufacturing process is becoming more automated and geographically concentrated.

Table 3.   Total capital expenditure of the main semiconductor producers of the United States, 1974-79
(Millions of dollars and percentage of total corporate revenue)

Firm	1974		1975		1976		1977		1978		1979[1]	
	Million $	%	Million $	%	Million $	%	Million $	%	Million $	%	Million $	%
Texas Instruments	148	9.4	66	4.8	136	8.2	199	9.7	300	11.8	350	12.0
Motorola	135	9.7	71	9.3	98	6.4	124	6.7	146	6.6	175	7.0
Fairchild Camera	41	10.4	21	7.0	36	8.0	22	4.7	32	6.0	70	11.4
Intel	13	9.6	11	8.0	32	14.2	45	15.9	90	22.6	100	17.9
Mostek	10	16.6	3	6.4	10	17.2	24	27.9	21	15.6	35	17.5

[1] Estimates.
Source: Ming Li: "The electronics industry: How the investment banker sees it", in *Speakers' Papers, Financial Times* conference, "Tomorrow in World Electronics", London, 21-22 March 1979, pp. 76-77.

The rising costs of some semiconductor manufacturers are affecting their liquidity and profit rates. This is leading to more intense industrial concentration through mergers and take-overs, and to a policy of forward integration towards manufacture of the finished product.[7] It could also lead to a slow-down in the pace of innovations. The increasing capital investment is shown in table 3.

The component sector is becoming even more capital-intensive than in the past, so that the proportion of direct labour cost in the total production cost is diminishing. This is particularly striking where automation of chip testing and assembly is introduced. The advantages of producing abroad are also being eroded for the same reason. This point will be discussed again in relation to developing countries.

Huge investment in research and development is essential in maintaining competitiveness in this fast-moving industry, and the speed at which technical obsolescence of products and processes sets in compels manufacturers to invest heavily in order to maintain their market position. Thus the expenditure on research and development by United States firms manufacturing semiconductors rose from $324 million in 1977 to $418 million in 1978. (Both the year-to-year increase in research and development expenditure and the amount spent on research and development in relation to sales remained steady, at 22-23 per cent and 8.5 per cent respectively.) However, whereas the figures for 1977 related to the 10 companies with sales in excess of 25 million dollars a year, the expenditure figures for 1978 related to the 11 companies with sales in excess of $35 million. These costs relate to improvements not only in semiconductors but also in materials, manufacturing techniques, design, applications and software.[8]

Other factors determining costs and prices are production volume and marketing policy. Evidence compiled on the basis of past experience in the semiconductor industry shows cost reductions of 25-30 per cent each time output doubles. The pricing policy of semiconductor firms appears to be to quote below actual cost, in anticipation of future reductions resulting from improved technology and expanded volume. Consequently, each firm tries to maintain or increase its share of the market by forgoing immediate profits in the hope of large profits when the product reaches maturity and few competitors are left in the field.[9]

In conclusion, industrial concentration in the components sector will tend to increase, accompanied by concentration of the technology. In order to amortise the high cost of developing and exploiting the technology, economies of scale will be sought through global diffusion. Thus, while the marketing strategy is one of world diffusion, manufacturing shows a different trend, namely that of geographical concentration in developed countries. Because the industry is heavily reliant on its technological edge over competitors, it is opposed to transfer of technology. This discourages possible de-concentration and a more widespread distribution of production in this key industry.

A similar situation already exists in the computer industry, with only a few companies controlling the business all over the world. At the same time, most of these companies are linked through marketing, technological and manufacturing joint-venture arrangements. This convergence is bound to intensify, along with business concentration due to the financial requirements of the industry, and the need to compete with the most advanced computers which are currently being marketed by IBM. The United States, Western Europe and Japan possess most of the world's computers, though the dominating force is the United States, which controls about 90 per cent of the computer sector in the market economy countries. All major computer manufacturers have plants in several countries and world-wide marketing and service networks. Table 4 shows the percentage geographical distribution of the world's computers. United States manufacturers supply almost 100 per cent of their home market, and it is only relatively recently that some inroads have been made by European and Japanese producers, particularly in respect of minicomputers and peripheral equipment. IBM alone supplies about 65 per cent of the United States market. The world market shares of the major computer manufacturers for mainframe computers, in terms of the value of total units installed, is as follows:[10]

	Per cent
International Business Machines	64.3
Honeywell	8.7
Sperry Rand-Univac	8.0
Burroughs	6.4
Control Data	4.1
National Cash Register	1.9
Others	6.6

Outside the United States, Japan is the country where local firms have the biggest share of the domestic market; this is due essentially to high protection in the past with the aim of developing local potential. In Western Europe some governments have followed a discriminatory purchasing policy for public sector equipment in an effort to encourage domestic industry and ensure uniformity. The share of IBM, and of United States firms generally, in the West European market has therefore been decreasing over the past few years. However, IBM may regain lost ground with its new generation of equipment (the 4300 series).

Table 4.   Percentage geographical distribution of computers, 1978-88

Countries or regions	1978	1983	1988
United States	44	44	40
Western Europe	24	25	26
Japan	10	8	8
Others	22	23	26

Source: Diebold Europe, 1979.

In developing countries the market share of IBM is about 70 per cent by value,[11] though smaller in terms of units. As other manufacturers enter the minicomputer market, IBM supremacy in terms of numbers will be eroded. A case in point is Brazil, where IBM had supplied 56 per cent of the number of mainframe and small computers in operation in 1978 but its share of the computer market went down to 17 per cent of the total if minicomputers were included in the reckoning. IBM has not been allowed so far to enter the minicomputer market in Brazil; on the other hand Olivetti, with no mainframe products, had captured 45 per cent of the minicomputer market, and supplied a total of 31 per cent of the number of computers in Brazil.[12]

IBM's policy of 100 per cent ownership of subsidiaries has also restricted its market penetration in several countries. An example is India: after assembling equipment there for many years, IBM preferred to withdraw in June 1978 rather than submit to partial local ownership. A state company took over servicing of the equipment. Similar restrictions have been encountered in other countries owing to requirements imposing partial local ownership and control in the long term. However, there is no doubt that the market supremacy of United States manufacturers, and their technical lead, will be maintained in the foreseeable future.

This supremacy is based on a solid and expanding home market, ensuring sales of $17,000 million in terms of computer installations and use in 1970, $42,000 million in 1978, and an expected $78,000 million in 1983. (Computer users' budgets include hardware, software services, supplies, communications, overheads and staff.)[13] The United States Department of Commerce, in its industrial survey of January 1979, predicted a growth in the country's computer industry shipments of 15 per cent in 1979, with an average 14 per cent annual increase up to 1985. Shipments in 1978 were worth $15,000 million. The computer industry was thought likely to remain the fastest growing sector of the United States manufacturing economy. For purposes of comparison it may be noted that other sectors were expected to grow as follows: metal-cutting machine tools, 8.8 per cent compound annual rate up to 1985; aerospace, 9.9 per cent; telephone, telegraph and electronic components, 9.2 per cent; photographic equipment and supplies, 9.3 per cent; plastics and resins, 8 per cent.[14]

The revenue of the main United States computer companies tripled in six years from $12,200 million in 1972 to $36,500 million in 1978. In 1978 IBM's share of the industry's total revenues was 57 per cent.[15]

As in the particular case of semiconductors already mentioned above research and development expenses in the United States information-processing industry in general run at about 6 per cent of sales, although the annual rate of increase of those expenses (16 to 18 per cent) is somewhat lower than for semiconductors alone, that being the particular product sector in which progress tends to be concentrated. In money terms, general research and development expenses rose from $1,995 million in 1977 (for the 25 computer firms with sales of more than $25 million a year only) to $2,539 million in 1978 (for the 36 firms in the information-processing industry with sales of more than $36 million a year each). In both years the expenses corresponded to 54-55 per cent of profits.[16] These tremendous costs compel the industry to follow an aggressive marketing and applications policy which in turn has an effect on the diffusion of information technology. Of the total outlay on research and development related to data processing, that of IBM alone accounted for 57 per cent in 1977 and 49 per cent in 1978.

The rapid expansion of the computer and information-processing field and its strategic industrial and security position have prompted rapid action in several countries.

Japan's rise as a consumer, producer and exporter of computer and related equipment has been tremendous. In 1960 Japanese computer production was valued at 2,000 million yen. This jumped to 142,000 million yen in 1968 and 415,000 million in 1976. By 1979 it was the second biggest market for such equipment among market economy countries.[17] This internal market expansion is coupled with an aggressive export effort supported by the Government. It was aimed to increase exports at an annual rate of 30 per cent up to 1985. By then, Japan expected to be exporting computer and related equipment worth $2,000 million a year.[18] The following are the export amounts (in millions of yen) for computers and peripherals:[19]

Item	1977	1982
Minicomputers	1	5
Computer systems	11	40
Peripherals	26	94

Of these Japanese exports, 35 per cent go to the United States and Canada, 25 per cent to Asia, 20 per cent to Western Europe, 4 per cent to the countries with centrally planned economies and 16 per cent to the rest of the world.

In Japan and Western Europe governments have invested heavily in the component, computer and telecommunications industries in an effort to maintain and develop national capabilities. The figures that follow are, for the most part, related to micro-electronics as such and do not reflect government support through other channels such as defence.

In Japan the Government, through the Ministry for International Trade and Industry, has launched a "breakthrough programme" with the big Japanese companies, which are organised into two research groups: Fujitsu-Hitachi-Mitsubishi and Nippon Electric Co.-Toshiba. From 1976 to 1979 the Government invested $150 million, the total programme amounting to more than $360 million including the companies' contributions. The aim was to develop chips that would be 60 times denser than existing ones,[20] and would be able to operate several times faster.

The countries of the European Communities have committed more than $2,200 million in support of micro-electronic technology. The United Kingdom is leading the field with $800 million for manufacturing facilities, applications and education; on top of this, the Government has allocated $95 million to several firms to perfect electron beam devices which can be used to increase the density of chips. A number of other schemes include support for ventures in visual display, word processing and telecommunications.

France has directly invested $500-600 million in support of micro-electronic technology, and the Government has assumed responsibility for increasing efficiency in the electronic sector. In the Federal Republic of Germany the Government allocated DM 1,574 million for the period 1976-79 to support the computer industry. Of this amount, 35 per cent was for industrial research and development and another 35 per cent for applications.[21]

The Government of Canada was to invest about C$50 million to help the country's electronics industry to become competitive, particularly in micro-electronics, and another C$20 million for communication satellites.[22]

On the continent of Europe a number of measures have already been taken. The European Communities have invested $86 million in a four-year programme (1976-80) to develop standards, procurement policies and other methods of maximising both the range of equipment and competition in the information industry. The European programme is also aimed at promoting co-operative development of new applications and joint industrial initiatives. Another joint project, "Eurodata", was started in 1977 by 17 European telecommunications authorities to study the needs of Europe's expanding data communications markets.[23]

This kind of supranational convergence in many fields, and the reinforcement of existing multinational companies, are due to the tremendous cost of research and of the development, manufacture and marketing of high technology products. The ultimate result of this convergence will depend on many factors, both economic and political. Discussion of the issues is already well under way. The situation in Europe has been described by the Director-General of the European Computing Services Association:

We either mean "owned and controlled" by Europeans; or we mean a much wider spectrum of criteria which includes investment, research, employment, design and management control.

The former definition would commit European nations to the rationalization of their manufacturing capability, for the scale of research and investment in an industry that must regard the whole world as its market (for that is what its competitors do) is too large for the present European companies on their own. The latter definition would diminish the continental confrontation, but would need to be accompanied by a social and business environment which ensures fair corporate behaviour in each country of operation by . . . European and non-European companies alike.[24]

There is considerable opposition to diminishing the intercontinental confrontation, particularly in cases in which issues of national independence are involved.[25] At the same time, the process of rationalising the European industry is taking place through a number of technical, marketing and joint-venture agreements, for example between Philips and Grundig, and between the Thomson-Brandt group and AEG-Telefunken. The Commission of the European Communities has prepared a comprehensive electronics development programme which was submitted to the Council of Ministers in November 1979.

The telecommunications industry is also dominated by multinational firms, despite the fact that in most countries this market is in the hands of the State. In 1973 the telecommunications industry in market economy countries had a total turnover of $22,000 million, 95 per cent of which was accounted for by 20 companies. At that time, for example, the telephone exchange market was controlled by only six companies.[26] The new generation of computer-controlled exchanges will further reinforce that trend. The main reason for this is that development of the equipment can cost up to $500 million and manufacturing is highly automated. To cover these tremendous costs and offset accelerating unemployment rates, export markets are essential. Even United States manufacturers with a large home market are pursuing a more aggressive export policy to supplement it. Some observers predict that by 1990 only four basic systems will be left in the field, one based in the United States, one in Japan and two in Europe;[27] this shows the highly competitive character of the market. At the same time, there is a high degree of concentration in machine services such as data banks and bases and software.[28] Despite rapid growth, the disparity in communications capabilities between developed and developing countries is considerable, and will remain so as communications in the developed countries become equated with telematics.

As telecommunication facilities in developing countries expand, the modernisation of the systems will inevitably evolve along the same lines as in the advanced countries. Indeed this is already evident in many countries, and a modern world-wide communications network is being created. The convergence of computer and telecommunications technology and further development of satellites and cheap earth stations will enhance the efficiency of this network.

The global scope of the information industry is an essential element in understanding the effects of the information revolution. The concentration of financial capital and technological knowledge in a few companies and countries will increase one-sided dependence rather than interdependence between countries. This point has caused concern in Western Europe and the developing countries for some time, and a number of plans have been made to counteract the trend.

Stressing the importance of global effects may seem superfluous. Up to now, analyses of dependence and interdependence have concentrated on questions of primary commodities, raw materials, trade and transfer of technology. However, we are entering a period in which the industrialisation prospects of the developing countries will be increasingly conditioned by the industrial strategy of the advanced countries, reinforced by current technological changes. The way in which this conditioning will take place will be described in Chapter 7, which deals with information technology and the developing countries.

Notes

[1] "World market annual survey", in *Electronics* (New York), 4 Jan.1979.

[2] ibid.

[3] For a discussion of the relations between high concentration and competition, particularly in regard to technological innovation, see Edmond Sciberras: *Multinational electronics companies and national economic policies*, New York University Graduate School of Business Administration, Contemporary Studies in Economic and Financial Analysis, Vol. 6 (Greenwich, Connecticut, JAI Press, 1977).

[4] Y. S. Chang: *The transfer of technology: Economies of offshore assembly, The case of semiconductor industry*, UNITAR Research Reports, No. 11 (New York, United Nations Institute for Training and Research, 1971), pp. 1, 17. See also United Nations Conference on Trade and Development: *International subcontracting arrangements in electronics between developed market-economy countries and developing countries*, Report by the UNCTAD secretariat (New York, United Nations, 1975), pp. 6–7.

[5] OECD: *Trade by commodities*, Series C, 1972 and 1976.

[6] Based on "World market annual survey", in *Electronics*, 5 Jan. 1978 and 4 Jan. 1979.

[7] Over the last decade or so, 21 semiconductor makers in the United States have been acquired, either wholly or in part, by large corporations in the United States or Europe. Fairchild Camera and Mostek have recently been taken over by Schlumberger and by United Technologies respectively. See "Peripherals: Mostek and the vanishing pioneers", in *Business Week* (New York), 15 Oct. 1979, p. 138.

[8] "R and D Scoreboard", 1977 and 1978 (*Business Week*, 3 July 1978 and 2 July 1979, based on data from Standard and Poor's Compustat Services Inc.).

[9] See on this point Chang, op. cit., and Sciberras, op. cit.

[10] Quantum Science Corporation, 1979. For details of market shares in Western Europe and Japan see *Datamation*, Sep. 1976, pp. 63 and 93 respectively.

[11] United Nations, Department of Economic and Social Affairs: *The application of computer technology for development*, Second report of the Secretary General (New York, 1973; Sales No.: E.73.II.A.12), p. 23.

[12] *Dati CAPRE* (Rio de Janeiro, Comissão Coordenadora das Atividades de Processamento Electronico), July 1978.

[13] International Data Corporation advertisement in *Fortune*, 5 June 1978.

[14] United States Department of Commerce: *Industrial outlook, 1979* (Washington, 1979).

[15] International Data Corporation advertisement in *Fortune*, 5 June 1978; and *Financial Times*, 6 Feb. 1979.

[16] See note 8.

[17] For details see Electronics Association of Japan: *Electronics in Japan 1978-79* (Tokyo, 1979), and Nihon Keiei Joho Kaihatsu Kyokai (Japan Computer Usage Development Institute, ed.): *Computer hakusho: Keiei joho system no kodoka to network no keisei, 1969* (White Paper,1969, on electronic computers: Advancement of management information system and development of data transmission networks; Tokyo, 1969); see also *Financial Times* survey, "Computer Industry", 19 Feb. 1979, p. VI.

[18] B.Uttal: "Exports won't come easy for Japan's computer industry", in *Fortune*, 9 Oct. 1978.

[19] United States Department of Commerce forecast in *Fortune*, 25 Sep. 1978.

[20] Uttal, op. cit.

[21] *New Scientist* (London), 11 Jan. 1979; Jean-Marie Fabre and Thierry Moulonguet: "L'industrie informatique: Développement, politiques et situations dans divers pays", Annex No. 7 in Nora and Minc: *L'informatisation de la société*, op. cit., *Annexes*, Vol. II: *Industries et services informatiques*, p. 61. For other investments in the field, see also *Financial Times*, 21 Dec. 1978.

[22] *International Herald Tribune* (Paris), 21-22 Apr. 1979, p. 9.

[23] "A programme for Europe", in *Computer Users' Yearbook, 1977* (Brighton, Sussex, and Bournemouth, Hampshire), p. 23.

[24] Quoted by J. Maisonrouge: *The information industry*, Seminar in La Hulpe, 31 Jan. 1979 (IBM; mimeographed).

[25] See Nora and Minc: *L'informatisation de la societe*, op. cit., particularly "Télématique et indépendance nationale", pp. 62-72.

[26] Nicolas Jéquier: *International technology transfers in the telecommunications industry*, OECD Development Centre, International meeting of researchers on the transfer of technology by

multinational firms, Conference paper No. 14 (Paris, 1975), p. 4, as cited in UNCTAD, Trade and Development Board, Committee on Transfer of Technology, Second Session (Geneva, . . . 1978): *Electronics in developing countries: Issues in transfer and development of technology*, Study by the UNCTAD secretariat in co-operation with Mr. Ashok Parthasarati, Secretary, Electronics Commission, Government of India, in his personal capacity (document TD/B/C.6/34, 12 Oct. 1978; mimeographed), p. 16.

[27] Max Wilkinson: "The day the French flew a phone exchange to Egypt", in *Financial Times*, 17 May 1979, p. 22.

[28] On this particular point, see "La guerre des données" in *Le Monde Diplomatique* (Paris), Nov. 1979, pp. 13-26, and Beca, "Les banques de données", op. cit., in Nora and Minc, op. cit., *Annexes*. With regard to software and services, see Francis Bacon and ICS Conseils: "Les sociétés de services et de conseils en informatique (SSCI)", Annex 9 in Nora and Minc, op. cit., *Annexes*, Vol. II: *Industrie et services informatiques*. See also "Le SICOB (XXXᵉ Salon International de l'Informatique, de la Communication et de l'Organisation de Bureau)", in *Le Monde* (Paris), 19 Sep. 1979, pp. 35-36; and *Le Monde*, 27 Sep. 1979, pp. 31-32.

APPLICATIONS AND SECTORAL EMPLOYMENT EFFECTS

4

As was to be expected in the light of the description of the technology in earlier chapters, the range of applications of micro-electronics is very extensive. The most common ones are listed at the end of the chapter. It will be seen that, of the products listed, some are new while others have simply been transformed; all of them are currently available. Scientific and technological research are excluded from the list, although that is the field in which applications are perhaps the most advanced. Most future applications are expected to alter old products (such as the pocket calculator) and processes rather than produce new ones. One of the reasons for this is that new products undergo a long maturing process, taking anything from five to ten years to reach a mass market.

It is unlikely that the diffusion of the technology will be felt as a sudden, explosive impact; rather, it will affect different sectors of the economy at varying paces. In the industrialised market economy countries, office work will be the most profoundly affected in the short run. Manufacturing will also be affected but to a lesser degree, because of its traditionally higher productivity. Applications in agriculture will be important but their social effect will be very slight. For the economy as a whole the most immediate effect will be on employment and international competitiveness. This chapter deals with applications and their effects on employment in the service and industrial sectors.

SERVICES

Office work

In the short run, in developed countries, the information revolution will have much the greatest socio-economic effect through the automation of offices, which will be introduced quickly for three main reasons. First, the sector is undercapitalised by comparision with manufacturing or agriculture: the typical office worker in advanced countries of the West is supported by only 2,000 dollars' worth of equipment, by comparison with 15 or 20 times that amount for factory workers. In the future it is expected that the office worker will be supported by equipment equivalent to five times the present value, and that most of it will be computer-based;[1] this increase in value will occur despite the fact that the price of computing power is decreasing by 20 to 30 per cent a year depending on the application. Secondly, the sector has shown low productivity increases. Increased office productivity will contribute to over-all company productivity and thus to its competitiveness; in some cases it will bring greater returns than equivalent increases in manufacturing operations. The lag in office productivity by comparison with other sectors is shown by the following percentage changes between 1969 and 1979:[2]

	Office staff	Industrial workers
Growth in numbers	45	6
Growth in productivity	4	80

Thirdly, office labour costs are high and rising, particularly for skilled labour, and the unionisation of office workers is increasing.

Micro-electronics intervenes in office work through its dual role as a production and organisational technology. Computerisation has already affected repetitive office work, which includes billing, accounting, routine clerical work and data processing in general. Past evidence shows a decrease in jobs or a loss of potential job creation as a result. Computerisation will be encouraged by the decreasing price of computers.

The most important new development as far as repetitive office work is concerned is the introduction of word processing equipment. The term "word processing" was introduced by IBM in 1964 to describe all automatic equipment that helped in the preparation of text. This rather wide definition covers everything from a memory typewriter that can be used on its own to shared-logic word processing systems in which a number of peripherals make use of the same computer. Word processors increase productivity enormously by reducing the amount of time secretaries spend retyping, correcting errors and handling papers. Productivity gains vary according to the type of application and equipment and the way in which the enterprise is organised. Evidence demonstrating the gains is abundant. A report of the Central Policy Review Staff of the United Kingdom concludes that the use of word processors instead of conventional typewriters results in fairly consistent productivity gains in excess of 100 per cent,[3] and a study by the Logica consulting firm, which compares traditional electric typewriters to word processors with television screens, shows productivity increases of more than 100 per cent in terms of additional words typed. In addition, typing is faster, and the system offers greater flexibility through operations such as automatic justification and standard paragraphs.[4] IBM found that the change to word processors among its 500 typists in the United Kingdom led to a productivity increase of 148 per cent.[5] A report by the Association of Professional, Executive, Clerical and Computer Staff (APEX) in the United Kingdom cites six cases in which word processors produced substantial productivity increases. In three of the cases there was a drastic reduction in staff and in one the workload was trebled without changes in manning levels; for the other two cases there were reports of productivity gains but no mention of changes in the number of employees.[6]

Further increases in productivity are obtained by the use of improved or "intelligent" dictation equipment, printers, photocopy machines, filing systems and automatic despatching. In order to take full advantage of the potential productivity increases, appropriate in-house and external communication systems will be essential.

The use of automatic equipment does not necessarily imply a reduction of staff, because in some cases new services will be provided or new features added. Nevertheless, as will be seen, it is estimated at present that there will be much displacement of labour in office work.

According to a report produced in the Federal Republic of Germany by Siemens, which covered 2.7 million office jobs, 43 per cent of the jobs could be standardised and 25-30 per cent automated. The potential varies according to sector; in public administration, 75 per cent of all jobs could be standardised and 38 per cent automated; in the private sector, possible savings of 25 to 38 per cent were suggested.[7] Some trade unions have estimated that automation to the degree described above could threaten the jobs of 2 million of the 5 million typists and secretaries in the Federal Republic.[8]

At the beginning of 1979 the average price ratio of standard word processors to electric typewriters in the Federal Republic was about 6:1 and the productivity ratio was 10:1. It is estimated that if the price ratio comes down to 5:1, a breakthrough can be expected.[9] In fact, present trends indicate that prices will be considerably reduced in the next two years.

The APEX report referred to above states that in offices where word processing is properly implemented it can produce among typing and secretarial staff a productivity improvement of 100 per cent. If it is assumed that this effect will be halved when the impact on related groups of electrical and administrative staff is considered, then the introduction of word processing into one-fifth of all United Kingdom offices by 1983 would mean, unless output increased to compensate, a fall of a quarter of a million in the number of office jobs.[10] As in the case of the Federal Republic of Germany, the introduction of word processors depends strictly on how much they cost by comparison with office workers. Figure 1 provides a rough indication of trends in the United Kingdom.

Micro-electronics also affects non-repetitive office work such as management, group decision-making and person-to-person business dealings. The changes here are related to increased independence from formal, labour-intensive information channels. The use of visual display units, facsimiles, teleconferencing and other forms of direct communication means that memoranda, messages and designs can be circulated instantly with little or no human intervention. Since data, word and graphic processing can be merged by the use of micro-electronic technology, independence from staff will be increased. The possibility of using terminals, word processors and facsimiles in combination opens the way to a completely different approach to handling information in the office, which will eliminate many administrative steps and functions.[11]

Public communications authorities in the industrialised countries are launching, or preparing to launch, national and international systems to meet the needs of the office sector. The United Kingdom Post Office expects to be able to offer the following services between 1980 and 1990: national and international centre-to-centre closed-circuit television, audio-visual telephones, cheap and

Figure 1. Trends in relative costs of word processors and typists in the United Kingdom, 1977-81

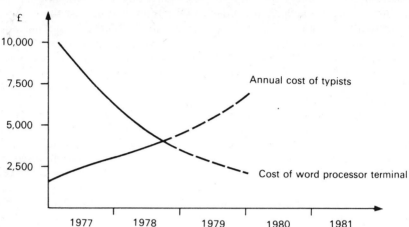

Source: Association of Professional, Executive, Clerical and Computer Staff: *Office technology: The trade union response*, First report of the APEX word processing working party (London, 1979), p. 18.

reliable facsimile terminals and telex, remote control of plants and machinery and remote reading of electricity and gas meters. Television displays of dialled data banks (Prestel) are already available. In the last decade of the century the full development of electronic mail is contemplated, along with home printing of electronically transmitted newspapers.[12] Indeed, the Post Office has already taken the first steps towards the introduction of electronic mail service which will become operational in late 1980. They will start leasing facsimile equipment early in the same year.[13] The French post office plans to install a million telecopier facsimile devices during the 1980s at a unit cost of about $300. They would be available for home or business use, probably under rental agreements as for the telephone. The authorities would like to convert this facsimile system into a full-scale electronic mail system, but are meeting with strong trade union opposition.[14] Among European countries the adoption of electronic mail techniques is expected to be quickest in the Federal Republic of Germany, which should have about 40,000 terminals by 1982 (as compared with 30,000 in the United Kingdom, for example). General projections estimate that 40 per cent of the 30 million letters handled daily in the United Kingdom could be delivered electronically.[15] The over-all expansion of electronic mail terminal equipment in Western Europe is expected to increase by 500 per cent between 1980 and 1987.[16] In the United States it is estimated that private communications satellites, using electronic mail techniques, could poach up to 30 per cent of the Post Office's volume of correspondence by 1986.[17] Related sectors, such as stationery, printing and office equipment manufacturers, will be similarly affected. A recent study predicts that, in the highly industrialised countries, about a million items of mail per day, which represents about 5 per cent of business-to-business post, will be transmitted electronically by 1982, and that this total will rise to about 3 million, representing 10 per cent of business-to-business post, by 1987. The effect on postal employment is not evaluated, but electronic mail will take away marginal business, thus affecting profits.[18]

Labour displacement in the office will primarily affect women workers either through a net job loss or through loss of job creation. Of all women at work in 1977 in the United Kingdom, 33.7 per cent were in secretarial and clerical jobs (as compared with 7 per cent in the case of men). Barron and Curnow argue that—

> With office automation, the sector of employment under threat is not, as in other cases of automation (data processing, factories, etc.), the general workforce, but the administrative force, with the emphasis predominantly on female workers; so any displacement would have major social consequences for female independence and family income.[19]

The same authors estimate that in the United Kingdom office automation will eventually affect to some degree the working lives of 2,150,000 people in clerical and administrative jobs.[20] The Nora-Minc report, commissioned by the French Government, reaches similar conclusions but provides no quantitative assessment.[21]

It is not possible to provide a detailed forecast of labour displacement in the office for a given country, and least of all for groups of countries. However, it is clear from the existing evidence that office automation will have a substantial effect on employment levels over the next four to eight years.

Printing

The printing industry, already affected by the introduction of electronic equipment, will also be affected by office automation and by electronic mail and newspapers. Major innovations in the printing industry have been electronic typesetting and photocomposition, which result in substantial productivity increases. The latest development is the use of photoscanning devices that allow the editor of a newspaper to see each page, complete with text, headlines and pictures, before it is transmitted via satellite to remote printing plants. From the evidence already available, it would seem that in the Federal Republic of Germany the number of printers and typesetters decreased by 30 per cent between 1970 and 1976.[22] In France, in 1977, 15,000 to 20,000 dismissals were being predicted in the industry by 1980.[23] In 1979 the London *Times* ceased publication for 11 months as a result of management-union conflict over the introduction of electronic typesetting, on which no final agreement has yet been reached.

Banking

Banking is a sector in which computer technology has long been used; it permits the creation of new services (such as credit cards and cheque cards) and greatly increases the efficiency of older ones. In France employment saving in this sector was estimated in 1978 at 30 per cent within a decade, not only in dismissals but also in the form of jobless growth. It was noted that already over the past two years, banks had cut back recruitment, whereas it had previously increased by 5 to 10 per cent a year.[24] In the United Kingdom employment in banking declined by over 50,000 between 1971 and 1976, while the value of deposit and other accounts rose from £27 million in 1972 to over £60 million in 1976.[25] The picture is similar in the Federal Republic of Germany where, for example, the number of jobs in the financial field declined by 6 per cent in 1976 despite an increase in the volume of business.[26] The centralisation of data processing, automatic cash tellers, electronic scanning of cheques and electronic transfer of funds will further reduce the labour force in this sector.

Other

Transport and wholesale and retail services are also being affected through rationalisation of transport systems, automatic or semi-automatic warehousing and stock control and "intelligent" cash-point terminals. In government, word processing and automation of a number of services has been added to the more traditional uses of computerised equipment for data processing, planning and forecasting. However, with regard to effects on employment, while a warning of possible redundancies in the social services owing to computerisation has been issued in France, the Central Policy Review Staff of the United Kingdom have concluded that ". . . the employment effect of computers in the Civil Service has been at best to restrain the growth of clerical employment and certainly not to reduce it."[27] This conclusion can be misleading: staffing levels are likely to be maintained, or even raised, with computer installations that use punched cards and thus require numerous workers to feed information into the system; however, punched-card systems are now being superseded, except for certain specific applications. It is clear that consideration of this type of change in computer technology is essential for the proper evaluation of past experience.[28]

New equipment in the service sector is often combined with organisational changes based on the concept of self-service. This is certainly not new; self-service restaurants and supermarkets are classic examples of this type of operation which has increased productivity enormously. However, the new equipment will increase the flexibility and sophistication of these services and of others, including vending machines, filling stations and home services such as laundering, cooking and sewing.

The array of applications of the new technology will reinforce the historical tendency of some service sectors towards concentration. In the United States the large retail companies' share of total sales increased from 30 per cent in 1948 to 44 per cent in 1972. This trend is expected to continue, causing unemployment in small businesses and shops.[29]

Cheap and widespread use of communication facilities could substantially alter education systems. Courses could be transmitted or bought off the shelf to be played in a videocassette system that could be used in conjunction with data banks and teachers or consultants. Such a system would effectively fulfil the following principles underlying the development of the Open University in the United Kingdom:

The characteristics of an open-learning system are that the learning opportunity is available to any person, at any level, at any time, in any place, and in any subject. . . . You should be able to do Chinese at three o'clock in the morning in a Welsh mountain village, or learn multiplication tables at 4 p.m. in central Birmingham. . . . An open learning system must be built on materials—print, tape, kits—and the student must be able to control his use of them. The teacher of course is still there. He is needed to man the base where the resources are, to run classes for those who want them, to run tutorials, to mark correspondence assignments, to make up courses from material banks and to provide a diagnostic and advisory service.[30]

It is possible to imagine a sort of "knowledge factory" producing kits in a variety of disciplines and offered in an "education supermarket".

Technological developments in medicine are slower, essentially because medicine deals with the least known of all "machines"—the human body and its complicated biological cycles. Progress in this field can be expected, but its pace is less predictable. Certainly "industrialisation" and "materialisation" are taking place and will be further developed by increases in health service productivity, particularly patient-staff ratios.

The view held by some that employment will continue to expand in the service sector in spite of micro-electronic technology should be carefully assessed. This section has given some idea of the wide range of applications in commercial, transport and public services, but many other services deal in information, and the need for more efficient management of information has been a major impetus behind the current information revolution. In addition, as has been pointed out, it may well be unsafe to extrapolate from trends in employment observed after the introduction of computers, partly because the new technology is making the use of computers more labour-saving too. Some services, such as tourism and entertainment, will undoubtedly increase their activities and possibly the numbers of their employees, but the question of whether such increases will compensate for losses elsewhere remains open.

INDUSTRY

The effect of micro-electronic applications on employment will be less dramatic in industry than in the service sector because of traditionally higher productivity levels. However, the substitution of a single, programmable chip for a range of separate electrical or mechanical processes has already considerably reduced labour input in many industries. Similarly, while the use of robots for hazardous work will improve worker safety, it will also contribute to labour displacement.

Industrial products themselves will be affected in a number of ways. Traditional products such as washing and sewing machines, cars and cash registers can be substantially improved by micro-electronic components. Articles such as watches and clocks can be completely transformed. Pocket calculators, word processors and television games will be cheaper or will offer better performance, and will thus find new markets.

Industrial performance will be improved by the application of micro-electronic technology to some products such as pollutant regulation equipment, to the rationalisation of administration and to the management of energy and resources.

In the following pages applications in selected industries and their effect on employment will be examined in more detail.

Electronics

Central to current changes is the electronics industry, which has enjoyed tremendous growth since its beginning. For example, in the United States, from 1963 to 1973, the electronics market grew by 202 per cent while gross national product per head grew by 70 per cent. In fact in most industrialised market economy countries, growth in this industry has persistently outstripped the increase in the gross national product per head.[31]

Along with the spectacular growth, labour displacement has occurred and technological change has greatly diminished job creation. As has been explained before (Chapter 3), the characteristics of the industry compel manufacturers to constantly cut costs, and labour has been a major component of total production cost.

However, some observers take the view that job losses in some areas will be offset by gains in others. These conflicting forecasts will be examined by taking examples of the employment situation in several branches of the electronics industry.

In the United Kingdom computer industry employment fell from 53,000 in 1971 to an estimated 43,000 in 1977, in spite of greatly increased sales. However, the labour force was expected to increase by about 5,000 by 1980. Job categories with the highest expected rate of growth included managers (23 per cent), scientists and technologists (21.3 per cent) and administrative and professional staff (19.9 per cent).[32]

Even in individual companies that have not yet completed the transition to production of electronic equipment, the percentage reduction in employment was as follows between 1969 and 1978:

National Cash Register	40
Olympia Werke	35
Alder Werke	20
Smith-Corona Machines	18
Olivetti	10

Production personnel as a percentage of total personnel fell as follows:

	1970	1978
Burroughs	44	31
National Cash Register	37	27
Nixdorf	38	22
Olivetti	45	31

The percentage of blue-collar workers decreased even more sharply. In the case of the Italian information processing industry, where there is still much mechanical production, the proportion of blue-collar workers in total personnel decreased from 53 to 38 per cent between 1970 and 1977 (from 64 to 47 per cent at Olivetti). At the same time, as operations become electronised, the percentage of research and development personnel increases in relation to total personnel. Between 1970 and 1978 research and development personnel increased from 6.4 to 10.1 per cent of the total at Burroughs and from 5.6 to 9 per cent at Olivetti.[33]

However, this increase clearly comes nowhere near offsetting labour displacement.

In the United States employment in the computer industry increased by 1 per cent between 1976 and 1977, while annual sales increased by 12.9 per cent, profits by 15.5 per cent and research and development expenditure by 16.3 per cent. The employment increase in this case was largely due to the development of relatively small computer companies (a point which will be taken up again below). However, in some of the bigger companies employment fell in 1977 (National Cash Register by 6.6 per cent, Honeywell by 2.4 per cent and Sperry Rand by 0.6 per cent).[34]

In the Federal Republic of Germany, out of 45 different manufacturing and mining industries, the manufacture of office and data-processing machines accounted for the third-highest production increase between 1970 and 1977, whereas employment in this industry fell by 20,600. Until 1976 the figures for unemployment in this industry were only slightly higher than for the others; the big change took place in 1977, when the number of persons employed was reduced by 2,300, or about 4 per cent, in spite of a rise of 27.8 per cent in the production index (1970 = 100) in the same year. It was in 1976-77 that micro-electronics began to be more generally used in the office equipment sector.[35]

One study argues that micro-electronics may create over 1 million jobs in Europe and the United States by 1987. These new jobs would be concentrated in the following four industries and their subsidiaries: consumer electronics, business communications, industrial electronic applications and the automotive industry. Of these jobs, about 60 per cent could be expected to be created in the United States and the rest in Europe.[36] An increase in employment can indeed be expected in firms producing electronic equipment and related products, though exceptions do occur as has just been shown with reference to office and data processing equipment in the Federal Republic of Germany. There is a belief that employment will be created as many new technologically based firms are founded to take advantage of developments in micro-electronics. This certainly did occur during the initial development of the electronic watch and clock industry, and also for microcomputers and peripherals. However, the original impetus was lost when the firms came up against the financial, marketing and research and development costs involved in staying in business. In the United States the number of microcomputer and home computer manufacturers has shrunk significantly since 1976. The conclusion of another study carried out in the United States is that the chances of success of a new product in a new market are 1 in 20.[37] Once small companies have created the market, the larger companies move in and capture it without risking any capital or other resources. This largely explains the high mortality rate of small and medium-sized new technology-based businesses and the likely trend for the future. The pervasiveness of micro-electronics will tend to attack traditional small-business product lines.

As might be expected, the competitive position of those sectors of [United Kingdom] industry already under threat from foreign competition is likely to be worsened by technical change based on microelectronics. There seems to be a dynamic of failure in which only successful firms can afford the high costs involved in adopting the new products and processes. Less successful firms thus become progressively less competitive.[38]

At the same time, automation is particularly justified in export industries because of economies of scale. These industries are already capital-intensive, and will become more so as the office automates. Eventually, the remaining labour-intensive activities in these industries will be centred on research and development, planning and management.

Furthermore, the reduction of employment in manufacturing is not necessarily offset by the expansion of market outlets such as those for pocket calculators, television games and data view systems, or by the creation of new products, such as the no-film camera, whose design often dispenses altogether with labour-intensive techniques. On present evidence, it seems that new products and processes do not substantially affect the general prospect of a reduction in employment levels:

It is hard to envisage new products, whether consumer or capital goods, which are of sufficient volume and value to absorb the labour displacement put forward [10–15 per cent or more of the workforce] at least in the same timescale. Studies of the labour-displacement effect of electronic watches versus mechanical watches, even of the apparently new product of the portable calculator versus the reduced demand for the electromechanical desk machines, suggest that the negative labour effect is much higher than the relatively trivial positive labour requirement for, say, electronic TV games.[39]

It is important to notice in this respect that in the information-processing industry the range of products has increased considerably but employment has decreased (the number of products in the Olivetti catalogue rose from 95 in 1965 to over 600 in 1977-78).[40] Although new products such as pocket calculators create new markets, they are also substitutes for older electromechanical devices.

Watches

A classic example of an industry deeply transformed by electronics is the watch and clock industry. The failure of the timepiece industries of Switzerland and the Federal Republic of Germany rapidly to incorporate electronic technology in the face of competition from Japan and the United States has led to substantial losses in employment and international competitiveness. The Swiss timepiece industry's share of the world market was reduced from 44 per cent of unit sales in 1970 to 29 per cent in 1977, and employment dropped by 40 per cent. In the clock industry of the Federal Republic of Germany employment was reduced from 32,000 to 18,000 or less in the 1970s.[41] The transition to electronics in this industry is not yet over. A recent study forecasts that the percentage of electronic, as opposed to traditional, watches will rise from 25 per cent in 1978 to 37 per cent in 1980 and 54 per cent in 1985. Similarly, electronic units accounted for 32 per cent of world clock production in 1978; this proportion was expected to rise to 55 per cent in 1980 and 63 per cent in 1985.[42] Given these trends, further reduction of the labour force can be expected, although not on the same scale as in the past. The transition to electronic watches has resulted in industrial concentration. Whereas traditionally the industry was made up of hundreds of small timepiece and component makers, it is now becoming increasingly dominated by nine major integrated manufacturers, three based in Japan, three in the United States, two in Switzerland and one in the USSR. Together, these producers account for nearly two-thirds of the world market in unit terms.[43] The Swiss industry, in particular, has undergone major restructuring. In the past, assemblers designed a model and then invited bids from component makers for various parts. This system has become obsolete and uneconomical, and many small companies have gone bankrupt. The combination of price cuts and the high costs of advertising and establishing expensive distribution networks combined to defeat many of the smaller companies that did attempt to use the new technology.

Cash registers

The National Cash Register Co. has been one of the companies most affected by the transition from electromechanical to electronic devices. Total employment in its manufacturing plants was reduced from 37,000 in 1970 to 18,000 in 1975.[44] With regard to products, the result has been the development of a highly flexible new line of point-of-sale equipment. Some of this equipment performs the functions of a computer terminal or is easily upgraded to become one, thus producing important changes in retailing. The cost of the transition was $169 million, plus heavy investment in retraining and an increase in the research and development allocation owing to continuing product obsolescence.[45]

Telephone switching equipment

The transition to electronic or semi-electronic equipment has had tremendous effects on the telecommunications industry and has also substantially increased productivity. In the United States Western Electric, the manufacturing arm of American Telephone and Telegraph, which supplies most of the telephone systems in North America, reduced its labour force from 39,000 in 1970 to 19,000 in 1976. The firm estimated that the introduction of new equipment would reduce the number of workers needed for repair, maintenance and installation work by 75 per cent. A report on orders placed by the United Kingdom Post Office for telephone equipment indicated that the manufacture of a million pounds' worth of electromechanical equipment would employ 160 people a year, compared with 80 people for the same value of semi-electronic equipment. In the United Kingdom the telephone supply industry workforce dropped from 91,000 in 1971 to 56,800 in 1977, and is being reduced even further. This takes into consideration the fact that 24 per cent of the orders placed by the Post Office for 1979-80 are for electomechanical equipment. The Swedish L. M. Ericsson company reduced its workforce from 15,300 in mid-1975 to 10,300 in mid-1978. By the time the change-over to electronic equipment is completed the company expects to have halved its labour force.[46]

Cars

The changes in the automobile industry tend to take the form of upgrading process instrumentation and control or introducing completely new electronically based systems. The car industry has moved quickly towards automation, mainly because of high labour costs and international and national competition. The product itself is changing through the use of integrated circuits for engine control and for improving the performance of the braking and electrical systems. One of the main features of the manufacture of this equipment is the rapid development of robots. The system used in Fiat's Robogate factory, for example, is designed to weld one car body per minute.

> This system has . . . been developed so that each self-propelled trolley . . . can be individually programmed to either of two vehicle types. First, the trolley is loaded with a jig, which carries references to a particular body type. It then starts its journey. The floor plan, body sides and roof are automatically assembled in succession on the jig. Sensors recognise the type of car to be assembled and ensure that the correct body parts are chosen. The robocarrier then moves to the various robot welding stations, electronically triggering the welders to carry out the operations relevant to the particular bodyshell. . . . Two different body types will be produced down each line, but this is not the limit of the system. Larger numbers of body types could be produced if required.
>
> The heart of the Robogate system is a computer; once programmed it controls the product mix [and] the . . . trolleys, ensures that the correct components are loaded aboard the trolleys, and finally carries out a series of quality control checks.
>
> The major breakthrough which the Robogate achieves is that, at the turn of a switch, the model mix can be altered . . . . [47]

The interesting aspect of the Fiat factory is its layout, which permits the flexible manufacturing system associated with batch production. In this system the product being made is carried by a pallet or trolley through various stations where automated units add components, drill holes, weld or perform other tasks.

Other automobile manufacturers are following suit:

- Volvo has installed 29 robots at a cost of £3.5 million, which perform the work previously done by a total of 70 men on two shifts;
- Peugeot has 15 robots for spot welding;
- British Leyland has installed two robot lines for the production of the new "Mini", to start at the end of 1980;
- Volkswagen, using robots, is producing the "Derby" model at a very low capital cost. It has about 100 robots working at present.[48]

Japan, the leader in this field, is believed to have 70,000 industrial robots at work, of which 20,000 are high-performance devices and one-third are used in the car industry. Developments in this field are fast. It was forecast that British industry would double the number of industrial robots from 100 to 200 high-performance programmable devices during 1979.[49] Some estimates of the United States robot market put current annual sales at $60 million, going up to $3,000 million by 1990. By then, robots could have displaced 5 per cent of the blue-collar labour force, particularly in the heaviest or most dangerous jobs. Unimation Inc., the world's largest manufacturer of industrial robots, estimates that a robot can work for less than half the hourly cost of labour in the United States.[50]

Other examples of the use of robots include the IBM mechanical assembler, which puts together eight parts of a typewriter in 45 seconds and can be adapted to a wide range of similar industries; the electronic robot used by the British Gas Corporation for inspection of pipelines;[51] the long-term development of "armchair" mining and automation at the coal face;[52] the use of robots in the diecasting industry for the unloading of machines, die cleaning and lubrication, and the loading of inserts and parts into trim presses.[53] Robots are being used to assemble the 17 parts of an alternator for commercial motor vehicles in

2 minutes 42 seconds; in a study of this robot various alternative costs were examined, and it was argued that once sufficiently developed, programmable devices would be economically attractive for product volumes as low as 300,000 units a year, and should remain cheaper than special-purpose machines up to volumes of 3 million units.[54]

Self-programming robots are also being developed. Here the machine, once it has been used in a prescribed manner and has stored the sequence in a computer memory, will repeat the sequence automatically.[55] These devices are currently being used in the car industry for painting, and in sewing machines, garment cutters and machine tools. Future developments in robots will depend on sensory devices and mechanical performance based on hydraulics and pneumatics. It is in these respects that the development of "artificial intelligence", which deals with machine reaction to non-programmed occurrences, will be most important.[56]

The applications cited above are part of the micro-electronic tidal wave; they are either in use already or will be very soon. However, increases in productivity are not only related to new equipment; some of the changes taking place in industry also reflect a radical approach to processes. It is now possible to automate small production runs for batch production and to produce more varied and individualised goods on the same production line.[57] There would be many uses for this type of automation in engineering, where about 75 per cent of all components are produced in batches of 50 and about 95 per cent of the time a product is in the workshop is spent in transport and waiting. In a traditional factory this time lapse denotes a waste of capital which could largely be eliminated in a computer-integrated automatic factory. The advent of more powerful microprocessors will permit the full development of what has been termed "distributed intelligence", which allows individualised and co-ordinated control of different operations within the same process.

OTHER FIELDS, AND SOME OVER-ALL EFFECTS

Two other fields of application should be briefly mentioned. The first is agriculture, where micro-electronics is being used in cattle monitoring, and computers are being used for calculating feeding formulas and fertiliser mixes, as well as for crop control. Through genetic manipulation, a number of agricultural products are being standardised, which will facilitate mechanisation and automation. The second field is the liberal professions. Architects and designers can employ computer-aided design techniques; computerised diagnostic equipment is available for medical use, and lawyers will be able to use terminals linked to data banks.[58]

Employment projections are tricky enough without considering technological displacement that is largely conjectural. Furthermore, it is extremely difficult to isolate micro-electronic from other agents active in automation and from changes in organisation. Nevertheless, in view of the nature of the technology, its persuasiveness and its speed of diffusion, there is ample evidence to support general agreement about its labour displacement effects. The argument is not whether jobs will be lost, but whether other jobs will be created elsewhere. This question is debatable, and the answers are largely a matter of speculation.

There are over-all estimates of effects on employment, job content and skills for some countries but, at this stage, they should only be taken as rough indicators of likely trends. It has been claimed that according to government sources 2.5 million jobs would be affected in the Federal Republic of Germany by developments in electronics in 1979 and 1980, not only through productivity increases but also through changes in job characteristics and skill requirements. In the long term, 45 to 50 per cent of all jobs would be both qualitatively and quantitatively modified.[59] In the United Kingdom predictions vary considerably. Barron and Curnow argue that should the labour displacement consequences take place in a society experiencing low growth and balance-of-payments problems it would probably face levels of unemployment of about 10-15 per cent or more of the labour force.[60] Stressing the difficulties of forecasting, the Central Policy Review Staff argues that predictions of 3.5 million unemployed as a result of the application of micro-electronic technology overestimate the speed of diffusion:

... Everything will depend on whether job creation matches job reduction. ... We have yet to be convinced that micro-electronics will be a major factor for the worse, unless the general prospects for employment make for increased unwillingness to accept technological change.[61]

In another study for the United Kingdom the reduction in the number of jobs is expected to total 5.2 million over 25 years; this is equivalent to 23.2 per cent of the reference labour force of 1978. Of the jobs in question, 4.6 per cent will have disappeared by the end of 1983, and up to 17 per cent by 1993.[62]

According to a report prepared by a committee organised under the sponsorship of the Japanese Ministry of International Trade and Industry—

... in view of slower economic growth and the inevitable changes in industrial structure that are foreseen in the 1980's, Japan will, sooner or later, face the problem of interaction between technological innovation and employment. In particular, if this country chooses to emphasize knowledge-intensive industries in the 1980's, it will not be able to avoid employment problems, arising from the extensive use of microelectronics.[63]

That assessment is based on the results of a survey of the employment effects of micro-electronics, according to different hypotheses of diffusion, up to 1985 and 1990. Starting from the total number of software engineers (80,310) in all industries in 1975, the authors foresee considerable job creation in the field of software engineering up to 1985, with an annual growth rate of between 15 and 35 per cent.[64] In the case of production workers engaged in the manufacture of general, electrical or precision machinery and equipment, the workforce should reach 2.43 million by natural increase in 1985. Depending on diffusion rates, the use of micro-electronics could reduce employment by between 200,000 and 480,000 jobs. With a maximum rate of diffusion, the real increase in the number of workers from 1975 to 1985 will be 70,000 (3.7 per cent); with the minimum rate of diffusion, the real increase over 1975 would be 340,000 (18.1 per cent).[65] A reduction of between 19,000 and 39,000 sales jobs in department stores (depending on the speed of diffusion of point-of-sale terminals) is forecast by 1990. If the diffusion rate is 100 per cent the reduction in sales staff from 1975 will amount in absolute terms to 4,000 persons (1.5 per cent); if the rate of diffusion is only 50 per cent there will be an absolute increase of 16,000 jobs

(6 per cent) over the 1975 level.[66] The report does not provide quantitative estimates for office automation but states that no substantial change in employment is expected.[67]

The information industry has so far kept a low profile on the employment question, emphasising instead the benefits and potentials of the technology:

> In the final analysis, the question is not whether the new technologies will or should be introduced, for they are clearly essential. We cannot manage effectively without them. The real question is how they should be introduced and developed. Here, there is much room for debate. Here is where labor, management, government and academia find their interests and responsibilities converging.[68]

Unskilled or semi-skilled jobs are not the only ones that will be affected; skilled manual and non-manual, as well as managerial jobs will be influenced both in number and in kind. The full consequences on employment will be felt only on completion of the new investment cycle in manufacturing and services, which is partly due to micro-electronic technology. Full-scale development of office automation, for instance, is expected to take place in the 1980s.

The question of where the new jobs will be created remains subject to much debate. What seems clear is that it is essential to adopt deliberate policies to ensure a smooth transition to a different system of production. C. Freeman argues that—

> Sometimes people tend to assume that these processes are automatically in balance, but it could be dangerous to over-simplify the process by which this balance is achieved. This is especially true of the major technical change now confronting us—the micro-processor revolution—which will both create many jobs and also destroy many jobs.[69]

The discussion in this section has been essentially concerned with employment effects. However, the magnitude of current and potential changes raises issues beyond employment. One of these issues is that micro-electronics seem to go hand in hand with business concentration as it reaches different sectors and applications. This produces changes in employment but also in the structure of the economy; small businesses and subcontracting arrangements are particularly vulnerable, and their collapse could lead to an erosion of competition at the national and international level.

Many other subjects that lie outside the allotted scope of this book need to be explored. These include changes in the internal structures of enterprises, in skill requirements, in work content and safety conditions and in salary structures, and the social assessment and control of a technology with great social implications.

APPENDIX. SOME OF THE MORE COMMON MICRO-ELECTRONIC APPLICATIONS[70]

Application	Examples
domestic appliances	washing machines, ovens, sewing machines, mixers
domestic regulators	central heating control, lighting control
leisure	television sets, radio, hi-fi equipment, audiovisual cameras, videocassettes

personal	pocket calculators, watches, teletex, personal computers
cars	dashboard displays, engine control (fuel supply, ignition, exhaust), collision avoidance, braking system, diagnostic system

Telecommunications

exchange equipment	public and private telephone exchanges
transmission	satellites, time-division multiplex transmission, radio, teleconferring, paging, electronic telephone, electronic mail and data transmission, electronic facsimiles, remote terminals, telex switching systems, teleprinters, electronic news gathering

Office

data processing	accounting, visible record computers
word processing	printers, word processors, copiers, facsimiles
filing	computer output and input microfilms, retrieval systems, recording, electronic archives
audio equipment	telephone answering machines, dictation, sales equipment

Trade

ordering	on-line ordering system
storage	automated warehousing, automatic stock control, inventory systems
distribution	computer-planned distribution networks
retail	cash control, point-of-sale equipment and terminals, automatic supermarkets
Banking	automatic cash tellers, electronic transfer of funds, credit card systems, cheque image processors

Printing

Printing	linotype electronic system, colour correction and storage, machine control

Computers and peripherals

minicomputers, microcomputers and memory equipment	magnetic disc or drum control, semiconductor memories, key punch systems, central processing units
input-output equipment	"intelligent" terminals, optical and laser character readers, printers and displays, front-end processors, multiplexors

Government and centralised services (private and public)

military and aerospace	air traffic control, radar system data processing, navigation systems, military communications, guided missiles, night viewing equipment, microwave blind landing, infra-red surveillance
education system	general educational techniques, audio-visual aids, individualised teaching
health system	filing, general hospital activities, body scanners and advanced diagnostic equipment, heart pacemakers, patient monitoring system, kidney dialysis equipment, computer-produced speech, electronic aids and sight for the blind
public control	centralised filing, police filing, traffic control
others	meteorology, pollution control

Manufacturing (general)

measuring and test equipment	devices for monitoring dimensions, temperature, weight and other factors
plant and personnel control	electronic clock-in, photocopying control, computer monitoring
robots (programmed and self-programming devices)	welders, carriers, painters, etc.
industrial processes—monitoring and control	nuclear, steel, high-risk flow-line
machines and equipment	batch processing control, machine tools, welding machines, electroplating, textile machines, materials handling, volume manufacturing, energy control, mail handling equipment, cutters, diecasting, furnaces and related
design and construction	computer-aided design, civil engineering and related design and equipment

Notes

[1] J. C. Peterschmitt: "The impact of the minicomputer", in *Speakers' Papers, Financial Times* conference, "Tomorrow in World Electronics", London, 21-22 March 1979, p. 149.

[2] Franco de Benedetti: "The impact of electronic technology in the office", in *Speakers' Papers, Financial Times* conference, op. cit., p. 130.

[3] United Kingdom, Central Policy Review Staff: *Social and employment implications of microelectronics* (1978; mimeographed), pp. 7-8.

⁴ Loose publicity insert in Logica Nederland BV: *Annual Review, 1977* (London).

⁵ "Word processing confusion", in *Financial Times*, 3 Apr. 1978.

⁶ Association of Professional, Executive, Clerical and Computer Staff (APEX): *Office technology: The trade union response*, First report of the APEX word processing working party (London, 1979), p. 16.

⁷ The study referred to *(Büro 1990)* was produced by Siemens in 1976. Although the results are still classified as confidential, they have been reported in several places. The figures here are based on the most detailed account, in Friedrichs, *A new dimension of technical change and automation*, op. cit., p. 9 (previous citation in chapter 2, note 7 of the present work).

⁸ The figures of possible labour displacement based on the Siemens report appear in *New Scientist*, 8 June 1978, p. 650.

⁹ Friedrichs, loc. cit.

¹⁰ APEX, *Office technology*, op. cit., p. 19.

¹¹ The changes here do not only refer to clerical staff; important changes will also take place in managerial practices and manning levels. See, for instance, David Clutterbuck: "The future of work", in *International Management*, Aug. 1979, pp. 16-19, and G. Tavernier: "The executive's turn to be automated", in idem, Sep. 1979, pp. 48-51.

¹² United Kingdom, Post Office Review Committee (Carter Committee) report (London, HM Stationery Office, 1977).

¹³ *New Scientist*, 7 Dec. 1978, p. 770.

¹⁴ *Electronics*, 5 Jan. 1978, p. 6E; for Japan, see *Electronics*, 13 Oct. 1978, and for the Federal Republic of Germany, see *Electronics*, 10 Nov. 1978.

¹⁵ *New Scientist*, 11 Jan. 1979, p. 96. Also see a report by Mackintosh Consultants in *Financial Times*, 13 Dec. 1978.

¹⁶ Study by Communications Studies and Planning in conjunction with Mackintosh Consultants, as cited in *New Scientist*, 7 Dec. 1978, p. 770.

¹⁷ *The Economist*, 21 Apr. 1979, p. 114.

¹⁸ *New Scientist*, 11 Jan. 1979, p. 96.

¹⁹ Barron and Curnow, *The future with microelectronics*, op. cit., p. 153.

²⁰ ibid., p. 151.

²¹ Nora and Minc, *L'informatisation de la société*, op. cit., pp. 35-38.

²² ILO: *Growth, structural change and manpower policy: The challenge of the 1980s*, Report of the Director-General, Third European Regional Conference, Geneva, . . . 1979, p. 96.

²³ CFDT: *Les dégâts du progrès* (Paris, 1977), p. 211.

²⁴ Nora and Minc, op. cit., p. 36.

²⁵ Conference of the Technical, Administrative and Supervisory Section of the Amalgamated Union of Engineering Workers (AUEW-TASS), London, 1978, Document C, p. 6.

²⁶ ILO, *Growth, structural change and manpower policy*, op. cit., p. 95.

²⁷ Nora and Minc, op. cit., pp. 36-37, and United Kingdom Central Policy Review Staff, *Social and employment implications of microelectronics*, op. cit., p. 7.

²⁸ For a critical examination of this particular point raised by the Central Policy Review Staff, see Colin Hines and Graham Searle: *Automatic unemployment*, A discussion of the impact of microelectronic technology on UK employment and the responses this demands (London, Earth Resources Research, 1979), p. 4.

²⁹ For details of this and other sectors, see United States Congress, House of Representatives Small-Business Sub-Committee (Washington, 1979).

³⁰ Richard Freeman, in *The Listener* (London, British Broadcasting Corporation), 14 Oct. 1976, as quoted by Jonathan Gershuny: *After industrial society: The emerging self-service economy* (London, Macmillan, 1978), p. 90.

³¹ J. Michael McLean: *The impact of the microelectronics industry on the structure of the Canadian economy*, Institute for Research and Public Policy, Occasional Paper No. 8 (1979), p. 8.

³² *Computer Users' Yearbook*, op. cit., *1978*, p. 22.

³³ B. Lamborghini: *The diffusion of microelectronics in industrial companies*, Paper submitted to the microelectronics conference held at Zandvoort in 1979 under the auspices of the International Social Science Council, European Coordination Centre for Research and Documentation in Social Sciences (Vienna); mimeographed, p. 7.

³⁴ "R and D Scoreboard, 1977", in *Business Week*, 3 July 1978, p. 63.

[35] Friedrichs, op. cit., pp. 5-6.

[36] Arthur D. Little Inc.: *The strategic impact of intelligent electronics in the US and Western Europe, 1977-1987*, Report by a study group led by Jerry Wasserman, as quoted by Max Wilkinson: "Micro-electronics may create 1 m more jobs", in *Financial Times*, 20 Mar. 1979, p. 7.

[37] Quoted by Maisonrouge, *The information industry*, op. cit., p. 10.

[38] J. M. McLean and H. J. Rush: *The impact of microelectronics on the UK*, A suggested classification and illustrative case studies, University of Sussex, Science Policy Research Unit, SPRU Occasional Paper Series, No. 7 (1978; mimeographed), p. 50.

[39] Barron and Curnow, *The future with microelectronics*, op. cit., pp. 199, 201.

[40] Lamborghini, op. cit., p. 10.

[41] See *Business Week*, 5 June 1978, p. 89; *Financial Times*, 12 Sep. 1978, and David Dangelmayer: "The job-killers of Germany", in *New Scientist*, 8 June 1978, p. 650.

[42] Study by Mackintosh Consultants reported in the *Financial Times*, 28 Mar. 1979.

[43] *Business Week*, 5 June 1978, p. 88.

[44] National Cash Register (NCR), 1975 Annual Report. See also the *Sunday Times* Business News, 17 Sep. 1978.

[45] National Cash Register (NCR), op. cit.

[46] Colin Hines: *The chips are down* (London, Earth Resources Research, 1978), as quoted in *New Scientist*, 8 June 1978, p. 665; Report of the United Kingdom Post Office to the Secretary of State for Industry (London, 1977; mimeographed); AUEW-TASS Conference, op. cit., p. 7, and *Financial Times*, 15 Dec. 1978.

[47] *Fiat News* (Brentford, Middlesex, Fiat Motor Company (U.K.) Ltd.), RGSS/BAP/544, 27 Oct. 1977, p. 2.

[48] *British Robot Association Newsletter* (Kempston, Bedfordshire), No. 1, Summer 1978.

[49] "Robots move ahead", in *Financial Times*, 30 Jan. 1979, p. 15.

[50] Unimation Inc., report (undated).

[51] See, for instance, *Financial Times*, 14 June and 28 Nov. 1978.

[52] United Kingdom, National Coal Board, Mining Research and Development Establishment: *Mining beyond 2000 A.D.*, as analysed in David Fishlock: "The NCB's dream of the next century: Extracting coal without miners . . . but watch the capital cost", in *Financial Times*, 30 Jan. 1979, p. 18.

[53] *Diecasting and Metal Moulding* (St. Albans, Hertfordshire, and Etchingham, Sussex), Jan.-Feb. 1976, Sep.-Oct. 1976, Mar.-Apr. 1978 and May-June 1978, *passim*.

[54] James L. Nevins and Daniel E. Whitney: "Computer-controlled assembly", in *Scientific American*, Feb. 1978, pp. 62-74.

[55] Cooley, "Contradictions of science and technology in the productive process", op. cit., pp. 74-75.

[56] George and Humphries (eds.), *The robots are coming*, op. cit.

[57] Nevins and Whitney, op. cit.; *New Scientist*, 2 Nov. 1978, p. 359.

[58] Computer Aided Design Centre (Cambridge): *Computer aided design: An appraisal of the present state of the art* (1978; mimeographed); *New Scientist*, 22 Feb. 1979, p. 558; Society for Computers and Law: *A national law library – the way ahead* (Oxford, 1979).

[59] V. Hauff, in *Die Zeit* (Hamburg), 13 Oct. 1978, p. 62.

[60] Barron and Curnow, op. cit., p. 201.

[61] Central Policy Review Staff, *Social and employment implications of microelectronics*, op. cit., pp. 4-5.

[62] Clive Henkins and Barrie Sherman: *The collapse of work* (London, Eyre Methuen, 1979), p. 123.

[63] *A fact-finding study on the impacts of microcomputers on employment*, Summary prepared by a working group on behalf of the Special Committee on the Impacts of Microelectronics on Employment (1979; mimeographed), Foreword.

[64] ibid., p. 24.

[65] ibid., pp. 26-28.

[66] ibid., p. 31.

[67] ibid., p. 37.

[68] Maisonrouge, op. cit., p. 10. See also Phillips: *Microelectronics: The pivot of an industrial revolution?*, July 1979.

[69] C. Freeman: *Government policies for industrial innovation*, J. D. Bernal Memorial Lecture, Birkbeck College, May 1978. Science Policy Research Unit (SPRU), University of Sussex, 1978.

[70] Based on McLean and Rush, *The impact of microelectronics in the UK*, op. cit., and other sources.

SPREAD OF THE NEW TECHNOLOGY

5

The spread of micro-electronic technology should be assessed not only in terms of time but also of the extent of penetration. Both aspects of the matter are examined in this chapter in respect of several industries. Factors conditioning diffusion, such as technical and economic characteristics, the competitive environment and socio-political acceptability, will be identified.

COMPUTERS AND DATA PROCESSING

The first commercially available computers (1950) were followed by the transistorised computer in 1959 and the integrated circuit computer in 1964. The September 1966 issue of *Scientific American* cautiously explained that in future computers "possibly micro-electronic circuits [would] be used not only as logic elements but also as memory elements".[1] Thirteen years after the commercialisation of the first computer, rental payments in the United States exceeded $1,000 million per annum[2] and by that time the major industrialised countries were already heavily dependent on computers. Computers and related equipment had very soon become essential to the functioning of many industries and services.

Statistics on the rate and extent of their diffusion, both currently and in the future, are often inconsistent because of lack of uniformity in definitions, the computerisation of traditional equipment and the inadequacy of the Standard International Trade Classification. For these reasons, table 5 is from a single source; other sources give different estimates. Market studies support the general trend shown in table 5, and indicate very rapid growth for computer sales in terms both of value and of number of units sold. One survey put 1978 world computer sales at over $20,000 million and predicted sales of over $24,000 million in 1979.[3] In terms of number of units the market is expected to grow even faster as hardware prices go down and capability increases. It may be noted that minicomputers, which did not exist in 1960, constituted 40 per cent of the total number in 1970, 60 per cent in 1973 and 70 per cent in 1978, and were expected to constitute 85 per cent in 1983 and 90 per cent in 1988.[4] In the minicomputer section of the United States market the number will grow from 60,000 in 1978 to

Table 5. Geographical distribution of computers and related equipment, 1960-88
(Number of computers and value of computers and related equipment in $ '000,000,000)

Countries	1960		1970		1973		1978		1983[1]		1988[1]	
United States	5 500	8.8	65 000	92.6	110 000	124.2	200 000	193.6	400 000	302.4	700 000	403.2
Western Europe	1 500	2.6	21 000	40.5	55 000	62.3	110 000	124.8	225 000	224.0	450 000	320.0
Japan	400	0.5	6 000	7.5	19 000	16.8	45 000	33.6	70 000	49.6	140 000	76.0
Other countries	1 600	0.8	18 000	9.6	46 000	22.4	95 000	72.0	205 000	128.0	460 000	195.2
Whole world	9 000	12.7	110 000	150.2	230 000	225.7	450 000	424.0	900 000	704.0	1 750 000	994.4

[1] Forecasts.
Source: Diebold Europe, 1979.

150,000 by 1982. In the microcomputer section the growth was expected to be even faster, from 20,000 in 1978 to 60,000 in 1980.[5] The number of computers for use in the home, which compete for some applications with more powerful computers, is expected to grow even faster. In the world as a whole, by the end of 1977, United States manufacturers had installed small business computers to a total value of $2,600 million.[6]

The United States data processing market, which includes computers and related equipment, is doubling every five years, and this trend is paralleled in Europe and Japan. In some sectors of the market the speed of diffusion is even faster. According to a recent study of Western Europe, the market for microcomputers will grow from $69 million in 1977 to over $800 million in 1986, an average annual growth rate of 32 per cent.[7] Manufacturers in the United Kingdom sold £415 million of computer equipment and services in 1973, and £900 million in 1976-77. At the same time, United Kingdom imports of equipment grew from £259 million in 1973 to £706 million in 1977. The computer service industry doubled its revenues in four years, to £253 million in 1976-77. By the end of 1977 there were 72,937 computers in the United Kingdom, 20,000 of·them being minicomputers and 4,663 mainframes.[8]

The computer population in developing countries is only a small fraction of the world total. According to table 5 the category "other countries", which includes developing countries, will increase its share of the total number of computers from 21 per cent in 1978 to 26 per cent in 1988; this will constitute an increase of almost five times in number and three times in value. However, the bulk of the growth in the "other countries" category will come from the centrally planned economy countries, not the developing countries. The former have a growing computer industry which they are under pressure to develop rapidly owing, in some cases, to labour shortages, and to the need to increase efficiency in management, forecasting and planning. Nevertheless, other authors forecast a rapid increase in the number of computers in developing countries as well. Estimates of the number of computers (several thousand) operating in developing countries were published by the United Nations in 1973 on the basis of replies received to questionnaires in 1971.[9] Since the report expressed doubts about the accuracy of some of the governments' replies to the questionnaires, table 6 updates the figures for selected countries, using other sources. The value of data processing equipment exported to developing countries from industrialised market economy countries rose from $314 million in 1972 to $934 million in 1976: the amount tripled in five years. At the same time, imports of such equipment by industrialised market economy countries from developing countries were valued at $31 million in 1972 and $282 million in 1976; despite this rapid growth the trade gap grew from $283 million in 1972 to $652 million in 1976.[10]

Micro-electronic technology is not diffused only through the increasing use of computers and related equipment as such. Micro-electronics also transform other capital goods, and they in turn influence manufacturing processes. This is the case for machine tools: microprocessors are rapidly being incorporated in them for control purposes, and machine control will be further developed by "distributed intelligence" within plants. According to one forecast on manufacturing technology, by 1986 the production of numerically controlled and

Table 6. Increase in the number of computers in some developing countries in the 1970s

Country	1971	Most recent information	Annual growth of GNP per head, 1960-76[1]
Algeria	63[2]	120[3]	1.7
Bolivia	6[2]	64[4]	2.3
Brazil	840[5]	6 641[6]	4.8
India	183[2]	420[7]	1.3
Iraq	7[2]	48[3]	3.6
Kuwait	17[2]	35[3]	− 3.0
Nigeria	30[2]	50[8]	3.5

Sources: [1] The World Bank (Washington): *World Development Report, 1978*, pp. 76-77. [2] United Nations, Department of Economic and Social Affairs: *The application of computer technology for development*, Second Report of the Secretary-General (New York, 1973; Sales No.: E.73.II.A.12), pp. 28-30. [3] 1978. *An-Nahar Arab Report and Memo* (Beirut, Middle East Economic Consultants), 4 December 1978. [4] 1978. Centro Nacional de Computación (CENACO), La Paz. [5] 1972. Comissão Coordenadora das Atividades de Processamento Electronico (CAPRE), Rio de Janeiro. [6] 1978. Same source as note 5. [7] 1977. Government of India, Department of Electronics: *Annual report, 1977-78* (New Delhi, 1978). [8] 1976. International Electronic Information Service (1977), *Computer Yearbook 1977* (Michigan).

computer-controlled machines will reach 50 per cent of the value of the total number of metal-cutting machine tools produced in the United States.[11] This trend is also evident in the machine tool industry in Japan and Western and Eastern Europe, which are the main suppliers of developing countries. There is no reason to believe that this type of technology will not be diffused in developing countries: on the contrary, all the evidence shows that in principle most nationals of the developing countries themselves wish to adopt the most advanced technology. A very important factor affecting the diffusion of the technology is undoubtedly government action: an increasing number of countries are adopting centralised systems for the processing of requests for import licences, the monitoring of applications of the technology and the purchase of equipment for government needs. These measures tend to cover only computers and related equipment; it is easy to see that it would be difficult to apply them to the less visible offspring of micro-electronics such as the components used in the machine tools described above. To some extent, therefore, the diffusion of the technology proceeds without the knowledge of the authorities.

In Argentina the reduction of import tariffs and import deposits in the past few years has produced rapid penetration of computers. By reducing financial requirements, the new arrangements have prompted the growth of small firms importing small computers. The situation is similar in Chile, where no regulatory policy exists, except for standardisation of computer languages and procedures as a means of establishing a network of interactive computers.

In Brazil, besides the reservation of sections of the market for nationally assembled computers, a government board sets national guidelines for developments in the field. Companies see the regulations as restricting the market considerably. However, the Brazilian computer market grew at the rate of about

35 per cent a year in 1975 and 1976. The rate then declined to 17 per cent in 1977 and 11 per cent in 1978, partly because of regulatory measures introduced by the board on imports of medium-sized computers and of minicomputers. These restrictions were aimed at preparing the market for nationally produced equipment. Once this equipment hits the market, it is expected that demand will increase. Paradoxically, the uncertainty in the small computer section of the market induced a large number of customers to invest in big computers, so that their share of the total computer market grew by 36 per cent in 1978. The board has estimated that computer equipment to the value of $1,700 million will have been installed between 1977 and 1982, of which 32 per cent will be minicomputers.[12]

India has some of the most elaborate policies and regulations governing the puchase and use of computers. Equipment is manufactured under licence, about 30 per cent of the components being imported, and foreign investment in this field is regulated. The cost of equipment, coupled with the regulatory measures, have severely restricted the market. Given the current and potential effects of information technology, it is likely that an increasing number of developing countries will follow the Indian example and regulate both imports and use.

Penetration of the technology is partly due to marketing drives by manufacturers who are under constant pressure to open new markets. An additional cause is the rapid technical obsolescence of hardware. This compels the manufacturers to sell or rent as many computers of a given generation as possible. They also need an outlet for the growing number of second-hand computers from industrialised countries. The fact that the developing countries usually lack extensive knowledge and experience in the information field makes it easy for some multinational corporations to offload their second-hand or obsolete products in this way.[13]

SEMICONDUCTORS AND SPECIALISED EQUIPMENT

The data available on semiconductors and specialised equipment for offices and services, industry and communications refer mainly to industrialised countries. Nevertheless, the control those countries exert over world trade, rules of competition and new products has clear implications regarding penetration of developing countries.

Semiconductors

Twelve segments of the semiconductor market accounted for sales worth $26 million in 1970, and are expected to reach $3,530 million by 1982. The segments referred to are relatively new: motor vehicle dashboards, citizen-band radio, pocket calculators, digital timepieces, stereo equipment, home computers, microwave oven controls, telephone handsets, minicomputers and "intelligent" terminals. The segments exclude mainframe computers, telecommunication switching equipment and transmission systems, military and space equipment, laboratory and test equipment, office products, industrial equipment and many electronic consumer products.[14] Table 7 shows the expansion of the world

Table 7.   Sales of semiconductors and integrated circuits, 1976-81[1]
(Millions of dollars and *annual percentage rate of increase*)

Year	World total[2]		Companies based in the United States[3]	
	Semiconductors	Integrated circuits	Semiconductors	Integrated circuits
1976	6 635.1	3 491.0	–	–
1977	7 229.2 *9*	4 190.5 *20*	3 857	2 463
1978	8 267.8 *14*	5 008.8 *20*	4 705 *22*	3 147 *28*
1979	9 200.6 *10*[a]	5 829.8 *16*[a]	5 123 *9*[a]	3 556 *13*[a]
1980	–	–	5 846 *14*[b]	4 191 *18*[b]
1981	–	–	6 740 *15*[b]	4 987 *19*[b]

[1] Estimates for 1979 and forecasts for later years.   [2] Source: Market survey in *Electronics* (New York), 4 Jan 1979.   [3] Source: United States Semiconductor Industry Association, as quoted in *Electronics*, 12 Oct. 1978

semiconductor market in relation to integrated circuits (the basis for present developments) and shipments of integrated circuits by companies based in the United States. It should be noted that the United States, Japan and Europe consume 95 per cent of world production. Certain types of integrated circuits show an even more impressive rate of growth. For example, from 1978 to 1979, sales of microprocessors increased by 37 per cent in the United States, 28 per cent in Japan and 30 per cent in Western Europe.[15]

The behaviour of the semiconductor market (particularly integrated circuits) shows that mass consumption of chips for consumer goods tends to slow down after the product has been on the market for a few years. Calculator and watch chips represented 8 per cent of the integrated circuit market in 1977 ($328.1 million) but were expected to account for only 6 per cent in 1979 ($345 million) In the case of calculator chips, the market was expected to decrease from $214.5 million annual sales in 1977 to $191.9 million in 1979 and even further in later years.[16] As a consequence of the behaviour of different segments of the semiconductor market, manufacturers are following three basic lines of development: forward integration to manufacture finished products rather than components alone, concentration on office automation which is expected to be the fastest growing market in the 1980s, and a search for new applications. The over-all effect of these tendencies will be a more rapid and extensive penetration of micro-electronic technology.

Office equipment and services

The expansion of electronic office and service equipment is closely related to the general expansion of computers but has its own dynamic. This is illustrated in table 8. Although development of the computer sector has already been discussed, it is relevant to point out here that the market for small computers is growing faster than the rest of the sector, which suggests that computer technology is penetrating small enterprises.

Word processing equipment is at the centre of office automation, and its

Table 8. Sales of electronic office and service equipment in the United States, Japan and Western Europe, 1977-82
(Millions of dollars and *percentage annual rate of increase*)

Kind of equipment	United States 1977	1978		1979¹		1982²	Japan 1977	1978		1979¹		Western Europe 1977	1978		1979¹	
Office³	5 109.0	5 730.0	*12*	6 387.0	*11*	8 292.0	732.5	830.0	*13*	1 010.0	*22*	2 253.7	2 610.9	*16*	2 977.6	*14*
Word processing	800.0	1 000.0	*25*	1 200.0	*20*	1 830.0	–	–		–		143.9	186.2	*29*	243.2	*30*
Data entry/output	1 557.1	1 821.0	*17*	2 194.0	*20*	3 627.6	302.5	422.8	*40*	474.9	*12*	795.3	886.9	*12*	986.5	*11*
Data storage	2 218.8	2 492.6	*12*	2 785.5	*12*	3 463.0	1 047.5	1 225.0	*17*	3 667.5	*12*	–	–		–	
Data terminal	1 163.0	1 448.0	*25*	1 773.0	*22*	3 116.0	616.7	751.5	*22*	1 204.3	*60*	806.6	1 014.7	*25*	1 205.5	*18*
Point-of-sale	328.0	368.0	*12*	414.0	*13*	597.0	80.0	92.0	*15*	105.8	*15*	179.5	212.5	*18*	241.4	*14*
Banking	175.0	177.0	*1*	192.0	*8*	259.0	–	–		–		–	–		–	
Data processing⁴	8 035.0	9 842.5	*22*	12 303.0	*25*	1 962.0	3 808.7	4 225.0	*11*	4 711.7	*12*	5 622.0	6 344.4	*13*	7 258.0	*19*

[1] Estimates. [2] Forecasts. [3] For the United States office equipment includes: word processing, non-consumer calculators, dictation, copying, office facsimiles, electronic typesetting, accounting, bookkeeping, printing and duplication. For Japan and Western Europe it includes: word processing, billing and accounting machines, office and scientific calculators, copying and dictation equipment. [4] Includes computers of all sizes.

Source: Market survey in *Electronics*, 4 Jan. 1979.

Table 9. Annual percentage growth in demand for point-of-sale and banking terminals, 1976-82

Kind of terminal	United States	Japan	Western Europe	Average
Point-of-sale	10	20	22	13.5
Banking	15	20	21	18

Source: Fabre and Moulonguet, "L'industrie informatique", op. cit., p. 84.

expansion is extremely rapid. In the United States the market doubled in four years, and similar developments are expected in Europe during the 1980s. No such expansion is expected in Japan, however, because of the characteristics of the language; the Government is sponsoring considerable efforts to overcome the difficulties. Word processing statistics vary enormously owing to differences of definition. In Europe in general, Mackintosh Consultants estimate that the total market for automatic typewriters will increase from $60 million in 1976 to $131 million in 1981.[17] In France, according to estimates by the Institut Rémy-Genton, there were 10,300 word processors in 1977, and there would be 25,000 by 1980.[18] Olivetti's estimates for the world as a whole in 1977 are as follows:[19]

	No. of units
United States	85 000
Federal Republic of Germany	9 800
France	3 100
United Kingdom	3 000
Other European countries	2 020

Although Olivetti's estimates are lower than those from certain other sources which maintain that the European pool in 1977 was around 100,000 units, they agree in the case of the United States figure.[20] In spite of the discrepancies among them, all these estimates show the speed at which the technology is penetrating the basically labour-intensive office sector.

Point-of-sale equipment, which could affect employment in the retail trade, shows equally impressive growth figures. Again there is a problem of definition, but the same trends probably apply to even the most sophisticated equipment. In the banking sector the use of automatic cash tellers and other devices which are substitutes for labour is growing rapidly. One French report, based on manufacturers' estimates, predicts the growth of point-of-sale and banking equipment as shown in table 9.

These figures do not differ very substantially from those of the *Electronics* market survey (see table 8), except for one large discrepancy in the banking figure for the United States. This can be partly explained by the fact that customers there are reluctant to use electronic transfer of funds at cash-teller points. The cheque is still seen by customers as the only proper authorisation for

Table 10.   Sales of machine tool controls in the United States, Japan and Western Europe,
1977-82
(Millions of dollars and *annual percentage change*)

Year	United States		Japan		Western Europe	
	Total		Of which microprocessor controls			
1977	263.5	.	128.0	.	100.0	.	131.9	.
1978	307.9	*17*	155.0	*21*	175.0	*75*	147.9	*12*
1979	339.5	*10*	177.5	*15*	250.0	*43*	171.8	*14*
1982	436.0	.	2 250.0	.	.	.	.	.

Source: Market survey in *Electronics*, 4 Jan. 1979.

payment. It is planned to overcome the problem by the use of image-processing devices (at present in the experimental stages) which "read" cheques and transform them into digital data, saving the cost of handling and transporting millions of pieces of paper.[21]

Filing and archive equipment is growing equally fast thanks to the combination of photographic and laser techniques and micro-electronics. In the United States the micrographic (computer output microfilm) market, which stood at $730 million in 1977, was expected to double by 1982. It is estimated that micrographics can save up to 50 per cent of the time normally spent by a secretary on filing.[22]

Industrial equipment

Most data processing equipment is used for both office and manufacturing functions; the equipment considered in this section, however, is used directly on the shop floor. It includes industrial microcomputers which, in the United States, will increase from 77,000 units in 1977 to 455,000 in 1980 and 3,050,000 in 1985.[23] It has been estimated that between 1976 and 1982 the number of industrial terminals will increase at an average annual rate of 20 per cent in the United States, 32 per cent in Japan and 31 per cent in Western Europe, the average rate for all those countries being 26 per cent.[24] Part of the growth in industrial terminals and minicomputers can be attributed to the penetration of small industry by microprocessor technology.[25] Table 10 shows expansion of the market for machine tool controls. As can be seen, this market is growing particularly quickly in Japan. The data for microprocessor-controlled machine tools have been included for the United States; they are not available for Europe and Japan. Development of this section of the market will speed up as automation is more widely applied, for example to batch production. In the developed countries competition is strong in the machine tool sector, with the result that traditional equipment is being improved. This has consequences for the developing countries, which obtain their machine tools mainly from the highly industrialised countries.

Table 11. Sales of data communication equipment, facsimile terminals and fibre optic communication systems in the United States, Japan and Western Europe, 1977-82 (Millions of dollars and *annual percentage change*)

Country and year	Data communication equipment		Facsimile terminals		Fibre optic communication systems	
United States:						
1977	865.0	.	97.0	.	11.0	.
1978	1 042.0	*20*	113.5	*17*	22.0	*100*
1979	1 244.0	*19*	132.0	*16*	42.0	*91*
1982	1 520.0	.	187.0	.	147.0	.
Japan:						
1977	137.0	.	161.6	.	1.8	.
1978	162.5	*19*	202.9	*26*	2.8	*55*
1979	181.5	*12*	250.6	*24*	6.3	*125*
Western Europe:						
1977	140.5	.	–		–	
1978	165.9	*18*	–		–	
1979	203.5	*23*	–		–	

Source: Market survey in *Electronics*, 4 Jan. 1979.

Communications

The increasing need for communications facilities is clearly reflected in table 11, which shows how this market has grown. Data communication equipment and facsimiles are essential for office automation and the information revolution. Fibre optic communication systems have been included; this is an embryonic market which is expected to expand in the 1980s. For many applications, fibre optics will eventually supersede copper cables. In the United States this market is expected to grow to $1,200 million by 1987, from $11 million in 1977.[26] The linking of word processors and other electronic systems with communication equipment will give a further impetus to the upward trend of the market. According to International Data Corporation, there were 10,000 word processors with communication capabilities in 1977; the figure is expected to be 100,000 by the early 1980s.[27] In a recent report it is estimated that the market for electronic mail equipment will grow from $350 million in 1978 to $1,400 million in 1987 in the United States and from $180 million in 1978 to $1,100 million in 1987 in Western Europe. The same study predicts that by 1987, word processors will also be able to transmit graphs and charts. By that year, such improved word processor terminals should account for 45 per cent of the electronic mail market in Western Europe; simple communicating word processors will account for 33 per cent of the total and facsimiles for 22 per cent. In the United States the figures will be 62 per cent, 26 per cent and 12 per cent respectively.[28] It is forecast that for the telecommunications equipment market as a whole, the value of shipments will nearly double from the 1970s to the 1980s, from just under $190,000 million to over $365,000 million.[29]

MAIN CONSTRAINTS ON DIFFUSION

At the beginning of this chapter it was indicated that the development of information technology is conditioned by factors that are social and political as well as technical and economic. In this section the main emphasis will be placed on socio-political pressures affecting the diffusion of information technology. In fact, while technical shortcomings do exist, they are only of very slight importance for the spread of the technology in the great majority of applications. No attempt will be made here to detail all the technical constraints on the market, but some of the more important will be briefly mentioned. More important in this respect is the "technological behaviour" of the manufacturers. Their marketing policy is often criticised on these grounds, and it can create long-term distortions in the market.[30]

From a general point of view, perhaps the single most important technical constraint is the inadequate telecommunications infrastructure and services in many countries and the tardiness of post office authorities in introducing innovations.

Economically, there are three important factors to be taken into consideration. The first is the need to amortise old equipment, i.e. to make full use of previous investment. A rapid scrapping of obsolete equipment is not economic, particularly in capital-intensive industries. This is perhaps the single most important factor slowing down the diffusion of micro-electronic technology. Because of the tremendous financial strain on companies that are forced to go electronic, management will prefer gradual to sudden upgrading whenever possible. However, this rule applies to the automation of manufacturing rather than in the office. The second factor is the hardware and operational costs of the equipment, and the third is the cost of software development. In industry the cost of introducing innovations at the product or process level can become prohibitive, especially for small businesses. In addition, it is sometimes necessary to devise costly back-up systems in case of failure of the equipment, power cuts or other anomalies. This is particularly important for enterprises or institutions which depend heavily on computerised devices for day-to-day operation. In many respects, dependence on computers has increased the vulnerability of society.

A final technical point is that operation of computers can entail legal problems which arise from the questionable validity of documents, signatures and so on, produced automatically, and the need to maintain hard-copy instead of electronic records. Problems also arise from mistakes and the effects of electronic transfer of funds and electronic mail on bank charges and commercial transactions in general.[31] All these problems need to be considered in the overall evaluation of the economic significance of computers. In Europe, for instance, it is estimated that the potential losses from 20 computer applications analysed could reach $4,000 million by 1988, for a number of reasons, the relative importance of which in percentage terms is as follows:[32]

Wrong or distorted data fed into the system, inappropriate use of results, etc.	42
Breakdowns and faulty functioning of the equipment	33
Poor working methods (erroneous programming, incorrect use of the equipment, etc.)	15
Fraud and sabotage	10

The major socio-political constraint is obviously the possible labour displacement effects of the technology. This issue is discussed in Chapters 4 and 6; in that connection two points will be mentioned here. First, information technology has so far affected a relatively small numbers of workers. The comparatively mild effects so far are partly a result of incomplete investment cycles in some sectors such as the automobile industry. Only now, in the early 1980s, will office automation start to accelerate and is likely to cause major labour displacement, as already pointed out. The workers displaced will be mainly women, at a time of increasing female and general white-collar militancy. In some countries family incomes will suffer. Secondly, a large portion of the labour force of the 1980s will be culturally different from that of the past 20 years. The generation now entering the labour force is in general better educated than older generations, with higher expectations, and also with a different set of values in relation to work. The question of job satisfaction will be crucial precisely when automation of the office becomes possible.

The demand for effective means of exerting social control over technologies with social effects is likely to increase in political and social importance. In general, it is probably true to say that, in the advanced countries of the West, the diffusion of any technology with significant social effects will not be smooth. The current struggle over nuclear technology immediately comes to mind.[33] Most public concern is centred on uses of information technology which may violate individual privacy. The sheer amount of personal information that is collected, the power recorded data acquires and the harm it can do if it is erroneous, the uncontrollable and automated way in which it is handled, give rise to some of the more frequently expressed misgivings. This has raised questions on the right of individuals to have access to data pertaining to them, the security measures needed to block access by unauthorised persons and the power and impersonality of huge private and public bureaucracies. A concern to safeguard privacy and confidentiality has caused a number of countries to pass legislation regulating the use of information stored in data banks and requiring manufacturers to provide special security measures for their systems. Several organisations such as the United Nations, the European Communities and the OECD are trying to reach agreement on an international convention and set of guidelines to harmonise legislation on the national and international flow of data (transborder data flows).[34]

Although the long-term effects of the technology and its cultural, social and political implications are a subject of extensive public debate in Europe, a debate which was partly sparked off by the Nora-Minc report, the outcome will affect the electronics market only if it is decided to avoid constituting an information society based on huge centralised and internationalised information systems and to aim instead at the establishment of more decentralised, individualised systems. It was argued in a recent French study that the development of telematics, i.e. the merger between communications and data processing, should be stopped, and that action should be taken to promote the large-scale use of independent microcomputers which would be operated on an individual basis to meet individual needs.[35] Although the development of such "mini-informatics" would affect national markets, it is doubtful whether a policy of this type will ever be implemented, because decisions in favour of telematics have already been taken by private enterprise and governments. A reversal of those decisions is unlikely because it would substantially erode the international competitiveness of the countries concerned. In fact among decision-makers the debate on information technology has been concentrated on how to apply the technology rather than on whether to use it at all.

Two of the most important short-term limitations on the expansion of the market are ignorance of the technology and lack of interest in its potential uses. The extent of the ignorance is reflected, for example, in the results of an industrial survey in the United Kingdom: it was found that 50 per cent of the firms surveyed were unaware of the potentialities of the technology, 45 per cent were aware of them but not using them and only 5 per cent were exploiting them.[36] European governments are moving fast to solve this problem through promotion, education and retraining, particularly in the engineering profession. The Government of the United Kingdom has committed about £40 million for educational purposes; one programme plans to reach 50,000 policy-makers. Under French Government plans about 10,000 microcomputers are to be installed in schools by 1984, and educational programmes have been undertaken in industry and the services.[37] These programmes, when completed, will have an effect on the extent of applications and on the size of the market. However, cultural and organisational patterns also need to be considered, particularly unwillingness to change traditional office or manufacturing practices. Automation is not, in general, a piecemeal change; nor does it consist simply in the addition of equipment: it generally involves a reorganisation of the work structure and the adoption of a systems approach, which is usually lacking, to office work. Similarly in the case of manufacturing, automation often involves changes in design, tooling, workload and skills, plus systematic research and development. The importance of traditional practices and behaviour should not be underestimated, even in companies which are aware of the possibilities of the technology. Simple technical and organisational measures could increase productivity considerably, but in a number of cases no such measures have been taken. This recalcitrance is not due to trade union action, as is sometimes supposed, but to archaic managerial approaches to productivity.

In developing countries the expansion of the market faces additional constraints, aside from government policy, mentioned earlier.

Data processing and related equipment has so far been exclusively designed to meet the needs of the industrially developed countries with high income levels, relatively expensive labour and a large industrial, service and government complex which requires and generates great quantities of data. Mass markets of this kind exist only in an embryonic form in newly industrialising countries. Thus current innovations are less relevant to their needs, and economies of scale cannot be exploited to the same extent: on the contrary, individual companies, in particular, need to spread the costs of infrastructure, service and operations over relatively few pieces of equipment, which raises costs. Existing computer installations in developing countries are seriously underutilised. A recent study reports that in Egypt the average computer was switched on only 65 per cent of the time, in Kuwait 63 per cent, in Saudi Arabia 37 per cent and in the Sudan a mere 27 per cent. In Brazil Average use was estimated at only 40 per cent of computing hours in the early 1970s. In Sri Lanka usage has been estimated to be as low as 30 per cent. Similar evidence exists for a number of other countries.[38]

One important factor that constrains the development of markets in developing countries is low labour costs, which make high-technology equipment less competitive. In such conditions amortisation also takes longer. This is one of the reasons for the rather slow diffusion of word processing in Argentina, Brazil and Venezuela. The lack of standardisation and of modern managerial practices has made software requirements more demanding and the total cost of installations accordingly more expensive. Ready-made software programmes are difficult to apply, even in banking. This situation has led, paradoxically, to considerable development of new software in some developing countries.

In many developing countries computers are often purchased on inadequate technical and economic grounds. Highly complex equipment is often used for tasks that could be performed by simpler machines; big mainframes are used where small processors would have been more than adequate, and comprehensive communications networks are installed where less sophisticated systems would have served. The technical obsolescence criteria of the developed countries are almost mechanically transmitted to the developing countries, and lead more often than not to unnecessary frequency in upgrading equipment. This is partly explained by the suppliers' unregulated dominance of the market.[39] Pressure is being brought to bear on multinational computer firms because of some of their practices. The case of IBM in India is one example, and restrictions on IBM and Burroughs have been imposed in Brazil. Long-term requirements for partial local ownership will also act as a regulatory measure on companies willing to submit to local regulations.

Nevertheless, when all the relevant factors are taken into account, it is evident that a rapid rate of diffusion in the developing countries can be foreseen, not only in the computer field but also in telecommunications and industrial applications. A slowdown can be expected in data and word processing, but the temptations of an array of "intelligent" equipment and the benefits that the technology can provide will make it difficult to enforce controls. This is not to say, however, that developing countries will be able to absorb micro-electronic technology fast enough to be able to apply it and compete with the industrialised countries; they lack the necessary information, skills and capital. Furthermore, if some European governments are having difficulties in convincing managers to use the technology, it is safe to assume that in many developing countries managers are almost completely unaware of its very existence. Lack of expertise in the computer field in developing countries will accentuate the uneconomic aspects of the use of computerised equipment. Although no empirical study exists, the losses are probably considerable, if European experience is anything to go by. However, the situation in Europe is different, since an important

Table 12. Annual percentage rates of growth of computer sales and gross domestic product in the United States, Japan and Western Europe, 1977-79[1]

Year and countries	Sales of computers,	Gross domestic product
1977		
United States	20	4.9
Japan	13	5.0
Western Europe	11	2.2
1978		
United States	18	3.8
Japan	16	5.8
Western Europe	14	2.7[1]
1979		
United States	19	3.9
Japan	17	4.6
Western Europe	14	3.2[1]

[1] Estimates and forecasts for more recent years. [2] These figures relate solely to States Members of the European Communities, and are taken from the *OECD Observer*, Jan. 1979.
Sources: For computers, *Electronics*, 5 Jan. 1978 and 4 Jan. 1979; for gross domestic product, unless indicated otherwise, *OECD Main Economic Indicators*, Oct. 1978.

proportion of the losses are occurring in respect of non-conventional or new applications that are not common in developing countries.

Trade unions have not significantly impeded adoption of the new technology, although there are examples of industrial action over computerisation in several countries. It would seem that most trade unions in developing countries do not treat this as a central issue in industrial relations.

By all standards, and in most of the industrialised countries, information technology is spreading at a rate faster than the general rate of growth of the economy. This can be clearly seen in table 12. Growth of the electronics industry will be accentuated by further technological changes, improvements in performance and reductions in relative price. Breakthroughs are imminent in many fields, particularly office automation. Growth and profit rates, coupled with the importance of the sector, are attracting investment and forcing the industry to maintain the pace of innovation and the search for new applications. This process will certainly increase the scope and pervasiveness of the technology while at the same time making it more economical. The relationship between the speed of diffusion and the time scale of employment changes is extremely difficult to establish because assimilation of the technology affects different sectors, subsectors and areas of work in different ways. It is safe to say, however, that the effect will be much more quickly felt than has been the case with regard to any previous labour-saving technologies. Many industries, offices and services in the industrialised countries will have changed substantially within five or ten years. As has been shown, some industries have been radically altered in the past four years by a decrease in direct labour input, and other industries such as the textile, garment, leather, footwear and automobile industries are beginning to follow.

Notes

[1] *Scientific American*, Sep. 1966, p. 65.

[2] Reid-Green,"A short history of computing", op. cit., (on page 12, note 8 of the present work), p. 90.

[3] Market survey in *Electronics*, 4 Jan. 1979.

[4] Diebold Europe, 1979.

[5] International Data Corporation advertisement in *Fortune*, 5 June 1978.

[6] International Data Corporation advertisement in *Fortune*, 8 Oct. 1978.

[7] Pactel: *The micro-computer market place in Western Europe 1978-'986* (1978).

[8] *Computer Users' Yearbook*, op. cit., *1974*, p. 22, and *1978*, pp. 24 and 26.

[9] United Nations, Department of Economic and Social Affairs: *The application of computer technology for development*, Second report of the Secretary-General (New York, 1973; Sales No.: E.73.II.A.12).

[10] OECD: *Trade by commodities*, Series C, 1972 and 1976.

[11] Society of Manufacturing Engineers and University of Michigan: *Delphi forecasts of manufacturing technology*, op. cit., p. 29.

[12] *Dati CAPRE*, (Rio de Janeiro, Comissão Coordenadora das Atividades de Processamento Electronico), July 1978.

[13] A detailed study of manufacturers' practices in India and other countries is to be found in Svein Erik Nilsen: *Use of computer technology in some developing countries*, Development Research and Action Programme, DERAP Working Paper 127 (Bergen, Chr. Michelsen Institute, 1978; for limited circulation only).

[14] C. Lester Hogan: "Is the integrated electronic industry facing demand saturation?", in *Speakers' Papers, Financial Times* conference, op. cit., p. 122.

[15] Market survey in *Electronics*, 4 Jan. 1979.

[16] Based on figures provided by market survey in *Electronics*.

[17] Mackintosh Consultants: *Yearbook of West European electronic data* (London, 1978).

[18] *Le Monde*, 19 Sep. 1978, p. 37.

[19] ibid., 23 Sep. 1978.

[20] *Financial Times* survey, 5 June 1978.

[21] R. Bennet and R. Strand: "Bankers and the cashless society", in *Business Horizon*, Vol. 21, No. 3, June 1978; *Financial Times*, 20 June 1978 and 23 Feb. 1979. See also Harvey L. Poppel: "The information revolution: Winners and losers", in *Harvard Business Review*, Jan.-Feb. 1978, pp. 14, 16, 159.

[22] International Data Corporation: "Productivity and information management", advertisement in *Fortune*, 12 Mar. 1979, pp. [44], [48].

[23] International Data Corporation advertisement in *Fortune*, 5 June 1978.

[24] Fabre and Moulonguet: "L'industrie informatique", op. cit., Annex No. 7 in Nora and Minc, *L'informatisation de la société*, op. cit., *Annexes*, Vol. II: *Industrie et services informatiques*, p. 84. A detailed study on the use of computers by small and medium-sized business in France, the Federal Republic of Germany and the Netherlands will be found in L. D. Neidleman: "Computer usage by small and medium sized European firms: An empirical study", in *Information and Management* (Amsterdam, International Federation for Information Processing, IAG Users' Group), Vol. 2, No. 2, May 1979, pp. 67-77.

[25] Fabre and Moulonguet, op. cit., pp. 3-97.

[26] "International Resource Development, Inc., 1977", in *Electronics*, 5 June 1978.

[27] International Data Corporation advertisement in *Fortune*, 8 Oct. 1978.

[28] Mackintosh Consultants' estimations, as reported in *Financial Times*, 12 Dec. 1978.

[29] Edgar A. Grabhorn and Alan B. Kamman: "Size and potential of the world's telecommunication markets", in ITU, *Telecommunication perspectives and economic implications*, op. cit., p. II.3.2.

[30] See for instance Barron and Curnow, *The future with microelectronics*, op. cit., especially Chapter 4.

[31] See on this point Philippe Lemoine: "Les problèmes juridiques soulevés par l'informatisation", document contributif nº 5, in Nora and Minc, op. cit., *Annexes*, Vol. IV: *Documents contributifs*, pp. 153-165.

[32] Association Internationale pour l'Etude de l'Economie de l'Assurance (Geneva): *Nature et importance des pertes économiques dans l'utilisation de l'informatique en Europe en 1988*, Etude menée . . . par Diebold Europe S.A., Paris, sous la direction de André George (1979), p. 36; see also a revised version, "Les risques de pertes indirectes induites par les systèmes informatiques", with English summary, in Association Internationale pour l'Etude de l'Economie de l'Assurance: *Studies in risk management*, Geneva Papers on Risk and Insurance, No. 13 (1979), pp. 56-80.

[33] See for instance K. Guild Nichols: "Technology on trial", in *OECD Observer* (Paris, Organisation for Economic Co-operation and Development), No. 98, May 1979, pp. 34-41.

[34] Numerous documents exist on this point. Particularly important is the OECD Information Computer Communications Policy Series, especially No. 1, *Transborder data flows and the protection of privacy*, op. cit. See also No. 2, *The use of international data networks in Europe* (Paris, 1979). In an earlier series – OECD Informatics Studies – see No. 10, *Policy issues in data protection and privacy*, Concepts and perspectives, Proceedings of the OECD seminar [on data protection and privacy, Paris], . . . 1974.

[35] Jean Bounine and Bruno Lussato: *Télématique ou privatique?*, Questions à Simon Nora et Alain Minc (Paris, Editions d'informatique, 1978).

[36] ACARD, *The applications of semiconductor technology*, op. cit., pp. 23-24. On this point see also Neidleman, op. cit., in note 24.

[37] *New Scientist*, 11 Jan. 1979, p. 96.

[38] "The computer as a development tool", in *An-Nahar Arab Report and Memo* (Beirut, Middle East Economic Consultants), 4 Dec. 1978; R. Ianuzzo: "Data processing in Brazil", in *Datamation*, May 1970; see also ILO: *Improving management's use of computers in developing countries*, Background paper for the second report of the Secretary-General to the General Assembly of the United Nations on the application of computers to development (Geneva, 1972; mimeographed).

[39] See for instance the study by Nilsen, op. cit.; also Christopher G. Baron: "Computers and employment in developing countries", in *International Labour Review* (Geneva, ILO), May-June 1976, pp. 329-344; ILO: *Improving management's use of computers in developing countries*, op. cit.; and "The computer as a development tool", op. cit.

INTERNATIONAL COMPETITION AND EMPLOYMENT

6

Industrialised market economy countries are being forced to adopt a deliberate policy of industrial readjustment. This policy is aimed at maintaining competitiveness, which is seen as one of the chief means of offsetting labour displacement caused by the new technology. Paradoxically, micro-electronic technology is emerging as one of the most important factors in maintaining competitiveness. The international economic background against which this readjustment policy is being worked out includes the lowest growth prospects for 25 years, high unemployment, a transition to new sources of energy, pressures for more equitable terms of trade from developing countries and the increasing capacity of some of those countries to export manufactured and semi-manufactured goods.

In the long industrialised market economy countries preservation of the lead in advanced technology products is, of course, a high priority. In addition, an attempt is being made to encourage the creation of firms based on new technology to exploit the potential of micro-electronics. These and already established companies will concentrate on improving existing products and processes and developing new ones. A number of financial and commercial measures are also being taken, in particular protectionist measures. The terms of world trade already favour the industrialised market economy countries and ensure diffusion of micro-electronic technology throughout the world. Table 13 shows the increased import and export share of manufactured goods of industrialised countries. The volume of these countries' exports increased by 8.5 per cent annually between 1960 and 1975;[1] developing countries were a major market in this period. In this way, the developed countries set the terms of competition and changes in manufactured goods are transferred, through trade, to the developing countries. It has been pointed out in Chapter 3 that there is a trend towards concentration of electronic product manufacture in the industrialised countries and a simultaneous drive to open even wider markets. This, added to the means being used by those countries to maintain competition, would appear to put them, collectively, in an unassailable position. However, the growth in manufactured exports from some developing countries is felt as a persistent warning prod in the side of the industrialised countries. Table 14 shows this growth, but it should be made clear that 45 per cent of the

Table 13. Percentage share of the increase in the trade in manufactures of industrialised countries,[1] by country group, 1960-75 (Current prices)

Trading partners	Imports	Exports
Within Western Europe	55	38
Other industrialised countries	34	24
Centrally planned economies[2]	2	6
Sub-total	91	68
Developing countries	9	29
Capital-surplus oil-exporting countries[3]	.	3
World	100	100

[1] Members of the Organisation for Economic Co-operation and Development, apart from Greece, Portugal, Spain and Turkey, which are classified as developing countries.   [2] Albania, Bulgaria, China, Cuba, Czechoslovakia, the German Democratic Republic, Hungary, the Democratic People's Republic of Korea, Mongolia, Poland, Romania and the USSR.   [3] Kuwait, Libyan Arab Jamahiriya, Oman, Qatar, Saudi Arabia and United Arab Emirates.
Source: The World Bank (Washington): *World Development Report, 1978*, table 12.

manufactured exports from developing countries come from only four countries (three of them in the Far East, the other being Spain).

A number of major national reports have appeared on the role of micro-electronic technology in maintaining competitiveness and employment. In the United Kingdom the Advisory Council for Applied Research and Development summarised the position by stating: ". . . if we neglect or reject [semiconductor technology] as a nation, the United Kingdom will join the ranks of the underdeveloped countries".[2] The Council also stated that because of the failure of the United Kingdom to respond adequately to technological changes British industry had already been overtaken by competitors in a number of areas such as food-processing equipment, cash registers, process instruments (except in cases in which overseas companies had used the technology in their United Kingdom subsidiaries), machine tools, telephone switching equipment and printing machinery. In many of these fields British industry had previously occupied a dominant position. A similar argument is used in the Nora-Minc report:

Today, for countries that have long been industrialised, productivity has become an over-riding constraint: these countries are simultaneously under pressure from the underindustrialised, overindustrialised and centrally planned economies.
This vice-like grip is beginning to tighten at a moment when—
■ the share of foreign trade has become essential to the national product and cannot be appreciably reduced without serious regression;
■ the political coalition of the developing countries is tending to upset the terms of trade in their favour; and
■ new technological breakthroughs make some countries hyper-competitive for future markets, while low labour costs reinforce the competition of less developed countries on traditional markets.
The result is that France is losing ground in a competitivity race which she entered for compelling reasons connected with foreign trade. The quest for productivity, from which

Table 14. Growth of merchandise exports, 1960-75
(1975 prices; average percentage rate per annum)

Item	World total	Industrialised countries	Developing countries
Food and beverages	4.1	5.2	2.8
Non-food agricultural products	4.5	5.6	2.6
Non-fuel minerals and metals	3.9	3.1	4.8
Fuel and energy	6.3	4.2	6.2
Manufactures	8.9	8.8	12.3
All merchandise	7.1	7.5	5.9

Source: The World Bank, *World Development Report, 1978*, op. cit., table 13.

the competition springs, has become an exogenous factor which conditions any domestic policy choice.[3]

A report issued under the auspices of the Science Policy Research Unit of the University of Sussex argues that "on balance the impact of microelectronics on both employment in the [United Kingdom] and the balance of payments will stem more from the slow rate of diffusion in many industries reducing international competitiveness, than from any direct effects".[4] The report reaches the following conclusions concerning the four case studies undertaken:

■ Materials handling: the ". . . tendency favouring the large chain retailer versus the small retailer has been in operation in Britain for the last twenty years; the impact of microelectronics could well be to further accelerate this tendency by increasing the relative efficiency of the large retail chains. The employment consequences for UK industry will, of course, be even more severe if the automated warehouse equipment is manufactured abroad; in either case the employment consequences of microelectronic innovation in the materials handling industry will not be limited."[5]

■ Textile machinery: ". . . there will be few spectacular examples of auto-mation using microelectronics in the textile industry itself but . . . the trend towards labour displacement will be continued and perhaps even accelerated."[6]

The report then warns against foreign competition in electronic and non-electronic innovations:

■ Motor vehicles: ". . . the potential for labour displacement due to product innovation, while not great, certainly does exist. Unemployment could conceivably be of even greater proportion if the industry's products fail to remain competitive. The major automobile manufacturers' attitude of 'proven technology' only may be a rational one considering the short term financial position; however, long term competitiveness would be affected if consumer preference favours the products of foreign firms which have incorporated microelectronics."[7]

- Word processing: ". . . a failure to adopt the new type of office equipment by UK employers could only lead to a further erosion of the competitiveness of British industry vis-à-vis its foreign competitors. The new microelectronic technology of word processing would thus seem to represent both an economic and a social threat of considerable dimensions.[8]

The Nora-Minc report reaches similar conclusions:

No doubt the massive productivity gains brought about by informatics, if they permanently improve competitivity, will reduce our external constraint. But even so they will not guarantee full employment. In order to overcompensate for the displacement of labour that a return to competitive capacity often requires, it would be necessary to increase external outlets to an extent that would seem implausible given the situation of the world market.[9]

These lengthy quotations emphasise two main points.

The first point concerns the apparent lack of choice for the market economy countries. Micro-electronic technology must be adopted, and goods and processes must be upgraded or created in order to maintain international competitiveness. The foregoing quotations constantly refer to the international aspects of the situation because productivity can no longer be manipulated within each domestic economy, being largely imposed from outside. This situation has resulted in a number of demands on social and economic planning and on trade unions. The latter face loss of jobs or industrial closures; accommodation to these problems has led to complex arrangements for hours of work, job security, and compensation agreements. Along with government and management, trade unions agree on the need to adopt the new technology. The trade unions' attitude towards current technological change can be summarised as "change by consent", although acute disputes have developed in a number of sectors. At the same time, most unions are stressing the need for long-term policies to ensure greater equity in sharing the wealth that will spring from the use of the technology. Union policies also emphasise the need to provide the services which will be required to meet social needs.[10]

The second point arising from the national reports quoted is that maintaining international competitiveness will not guarantee full employment but will only reduce the labour displacement. A discussion of this subject will be found in the next section of this chapter.

We have, then, two simultaneous and contradictory tendencies. On the one hand, fear of unemployment could slow down the diffusion of the technology; on the other, a failure to adopt the technology could create unemployment. Government and private industry have concentrated on meeting the second threat, and have invested heavily and taken quick action to promote and manufacture micro-electronics and to train the necessary personnel.

EMPLOYMENT AND UNEMPLOYMENT TRENDS

A range of economic, social and political prospects condition the response of developed countries to information technology, but the over-riding concern has been with the effect of information technology on employment as the most immediate, short-term effect. The advent of micro-electronics must be seen in the context of high levels of unemployment in the industrialised market economy countries. In 1960 unemployment averaged 3 per cent, and by 1969 it was down to 2.5 per cent. By 1979 it had steadily climbed again to an average of around 5.25 per cent, as a result of economic recession, energy prices, international competition and inflation. Some of the effects of the application of micro-electronic technology on employment were discussed in Chapter 4; the over-all conclusion reached there is in agreement with the following statement from Professor C. Freeman:

> I would be the first to agree that . . . the employment effects of the micro-processor revolution cannot be forecast. . . . But what has happened on a small scale already in the 1970s seems to me sufficiently indicative to warrant serious concern. If you take such areas as telecommunications, machine shops, automobile assembly, automated warehousing, printing and publishing, and clocks and watches, then there is already sufficient evidence available . . . to show that the labour-displacing consequences may be very severe indeed.[11]

Prospects for massive employment creation in the future are bleak owing to economic, demographic and other factors that affect developed and developing countries alike. The developed countries entered the 1980s with an average rate of unemployment of not much under 6 per cent. The *Prognos Euro-Report 1979* predicts that 12 million people will be unemployed in Western Europe by 1990 if no decisive changes in working hours and work organisation take place within the next decade. On average, the rate of potential unemployment would rise to 9 per cent in Western Europe. Although the effect of computerisation is recognised, the report expects new white-collar jobs to be created for researchers, programmers, documentalists and data processing personnel. These predictions are based on a real growth in gross national product of 3 per cent for the European Community countries up to 1990 and a parallel rate of growth in productivity. The report predicts an average of 8 per cent unemployment for European Community countries by 1983, or over 10 million people unemployed (from about 6 million in 1978).[12] However, much detailed research will be needed to advance from conjecture to fact.

Together with high unemployment prospects, the following three trends must be taken into account in order to clarify where technological change is likely to be most felt:

1. The long-term shift from agriculture to industry and services in the developed countries.[13] In the developing countries this shift seems to have been in the direction of the service rather than the industrial sector, although the movements do overlap.[14]

2. The increasing proportion of white-collar workers in the total labour force in all advanced countries. A similar trend is noticeable in developing countries. Thus an increasing number of office workers are classified as engaged in "manufacturing".

3. The incorporation of women into the labour force and the increasing unemployment of young workers.

Table 15 gives an over-all view of the shift from agriculture to services on a world-wide scale. The figures should be treated as indicators of orders of magnitude, because of the degree of aggregation. From the table it is clear that, in more developed regions, the fastest growing sector is services, with a decline in

Table 15. Labour force in agriculture, industry and services, 1950-70
(In thousands and *percentages of the total labour force*)

Regions[1]	Year	Agriculture		Industry		Services	
More developed	1950	149 330	*37.6*	120 748	*30.4*	127 358	*32.0*
	1960	124 336	*28.1*	152 732	*34.6*	164 730	*37.3*
	1970	89 319	*18.3*	183 477	*37.6*	215 134	*44.1*
Less developed	1950	558 390	*79.5*	58 921	*8.4*	85 403	*12.2*
	1960	623 988	*72.9*	108 607	*12.7*	123 007	*14.4*
	1970	679 372	*66.6*	162 550	*15.9*	178 762	*17.5*

[1] Designations are in conformity with United Nations practice:

More developed regions: Japan, Southern Africa, temperate South America, Eastern Europe, Northern Europe, Southern Europe, Western Europe, Northern America, Australia, New Zealand and USSR.

Less developed regions: China, Eastern South Asia, Middle South Asia, Western South Asia, Eastern Africa, Middle Africa, Northern Africa, Western Africa, Caribbean, Middle America mainland, tropical South America, Melanesia, Polynesia and Micronesia.

Source: ILO: *Labour force estimates and projections, 1950-2000* (second edition, Geneva, 1977), Vol. V: *World summary*, table 3, p. 40.

agriculture since 1950. In the case of the less developed regions, a similar pace of growth in employment is seen in industry and services with a slowdown of the speed of job creation in agriculture. The similar pace of growth in industry and services in developing countries can be largely explained by their late in-dustrialisation but the trend observed for the older industrialised countries is bound to be reproduced in the developing countries owing to the types of technology and development models used.

The more detailed figures provided in tables 16 and 17 further substantiate this point. The developed countries (table 16) all show a decrease in agricultural employment and a slow-down in job creation in industry since 1970. The figures for three of them (France, the Federal Republic of Germany and the United Kingdom) show an absolute decrease in the number of people employed in industry since 1970, thus accentuating the historical trend since between 1970 and 1975 their total labour forces showed net increases of 1.4 million, 1.15 million and 324,000 respectively. In the case of the United States the relative size of the industrial labour force has been shrinking since 1950 and the pace of job creation decreased significantly. Between 1970 and 1977 industry created 411,000 new jobs while the net increase in the labour force between 1970 and 1975 was 7.5 million.

Of the developing countries in table 17, Argentina, Bolivia and Chile show a decrease in the relative size of the industrial labour force. In the case of Bolivia there is a sharp absolute decrease since 1970 as well; this should, however, be assessed cautiously since it might be related to the size of sample and methodology used in the tabulation of the census or to inflated figures for 1970. Argentina and Chile have an industrial base which is relatively old by Latin American standards, and there are clear signs that job creation in industry is slowing down. Argentinian and Chilean industry created 40,000 and 48,000 jobs

Table 16.   Labour force in agriculture, industry and services in France, the Federal Republic of Germany, the United Kingdom and the United States, 1950 to the late 1970s (In thousands and *percentages of total labour force*)

Country	Year	Agriculture		Industry		Services	
France[1]	1950	6 102	*31.0*	6 806	*34.5*	6 797	*34.5*
	1960	4 345	*22.1*	7 606	*38.7*	7 710	*39.2*
	1970	2 871	*13.7*	8 301	*39.7*	9 765	*46.6*
	1975	2 099	*10.0*	8 074	*38.6*	10 770	*51.4*
Federal Republic of Germany[2]	1950	5 342	*23.0*	9 909	*43.0*	7 886	*34.0*
	1960	3 719	*14.2*	12 526	*47.7*	10 013	*38.1*
	1970	2 000	*7.5*	12 894	*48.0*	11 923	*44.5*
	1977	1 666	*6.4*	11 500	*44.5*	12 696	*49.0*
United Kingdom[3]	1950	1 275	*5.4*	11 655	*49.2*	10 741	*45.4*
	1960	974	*4.0*	11 581	*47.7*	11 706	*48.3*
	1970	721	*2.8*	11 586	*45.0*	13 422	*52.2*
	1976	660	*2.7*	9 756	*39.5*	14 297	*57.9*
United States[4]	1950	7 975	*12.2*	23 868	*36.6*	33 432	*51.2*
	1960	4 819	*6.6*	26 621	*36.4*	41 719	*57.0*
	1970	3 201	*3.7*	29 973	*34.4*	53 950	*61.9*
	1977	3 580	*3.6*	30 384	*30.8*	64 630	*65.6*

[1] 1975: Figures based on a 5 per cent sample tabulation of census return. Unemployed excluded.   [2] 1977: Official estimates. Persons seeking employment for the first time are excluded.   [3] 1976: Official mid-year estimates. Unemployed excluded.   [4] 1977: Estimates based on the results of labour force sample surveys. Persons seeking employment for the first time are excluded.

Source: For 1950, 1960 and 1970: ILO: *Labour force estimates and projections, 1950-2000*, op. cit., Vol. IV: *Northern America, Europe, Oceania and USSR*, table 3. For other dates: ILO: *Year Book of Labour Statistics, 1978* (Geneva, 1978), table 2A.

respectively in the 1960s. This has to be seen in the context of a net increase in the labour force of 1 million for Argentina and 380,000 for Chile. In Brazil the industrial share of the labour force increased. No fewer than 2.1 million new jobs were created in industry in ten years, with a net increase of the labour force of 7.1 million for the same period. This pace of growth is not expected to continue, however, particularly after the 1974 recession. Only in Brazil has the agricultural labour force increased, although not in relative terms.

In all the countries in the tables the service sector is growing more rapidly than the others, although it lags behind in terms of productivity. Coupled with the growth of the service sector of the economy, a change in the composition of the labour force has taken place, particularly in developed countries and to a lesser extent in developing countries.

Table 18 shows the change in three traditional categories of the labour force in four developed countries. The relative position of production workers, who could roughly be described as "blue-collar", is more or less constant; the proportion of clerical and managerial workers increased sharply while the proportion of agricultural and related workers decreased sharply. The combination of these two trends indicates that what scope there is for employment creation lies in the service sector, especially in clerical and managerial work. It is

Table 17. Labour force in agriculture, industry and services in Argentina, Bolivia, Brazil and Chile, 1950 to the 1970s
(In thousands and *percentages of total labour force*)

Country	Year	Agriculture		Industry		Services	
Argentina	1950	1 778	*25.1*	2 229	*31.5*	3 057	*43.2*
	1960	1 625	*19.9*	2 918	*35.8*	3 590	*44.1*
	1970	1 507	*16.3*	2 958	*32.1*	4 751	*51.5*
Bolivia[1]	1950	646	*61.4*	244	*23.1*	162	*15.4*
	1960	785	*60.9*	234	*18.1*	269	*20.9*
	1970	880	*55.4*	334	*21.0*	372	*23.4*
	1976	675	*47.0*	280	*19.5*	478	*33.3*
Brazil	1950	10 624	*59.7*	2 528	*14.2*	4 632	*26.0*
	1960	11 867	*51.8*	3 395	*14.8*	7 618	*33.3*
	1970	13 706	*45.6*	5 510	*18.3*	10 827	*36.0*
Chile	1950	707	*32.5*	668	*30.7*	796	*36.6*
	1960	751	*29.9*	758	*30.2*	996	*39.7*
	1970	685	*23.7*	806	*27.9*	1 393	*48.2*

[1] 1976 figures based on a 3.3 per cent sample tabulation of census returns. Persons classified under "activities not adequately described" are excluded.
Source: For 1950, 1960 and 1970: ILO: *Labour force estimates and projections, 1950-2000*, op. cit., Vol. III: *Latin America*, table 3. For Bolivia, 1976: ILO: *Year Book of Labour Statistics, 1978*, op. cit., table 2a.

precisely this sector, however, which will be most severely affected by information technology and its offspring, office automation. It is also within this sector that much of the female labour force is concentrated in developed countries, and increasingly in developing countries. Female activity rates have increased in most countries since 1950. Because women's potential participation in the labour force is not reflected in labour statistics, the actual size of the potential labour force is probably understated.

The poor prospects for large-scale job creation have important social consequences, particularly for the young. In the industrialised market economy countries as a whole (excluding Turkey) 6.9 million young workers between 15 and 25 years of age, or 10.7 per cent of the total, were unemployed in 1976. They accounted for 44 per cent of all the unemployed, whereas in 1970 the corresponding figure had been about 5 per cent. Not only are there more young people unemployed but the duration of their unemployment is increasing.[15] Table 19 provides a general view of past and future net increases in the labour force and current levels of unemployment in selected developed countries. From these figures it is easy to see that the situation looks grim. To absorb the net increase in the labour force in the 1980s, 11 million jobs will need to be created in the United States on the assumption that the decade begins with an unemployment level of 6 million. In the United Kingdom as many additional jobs will need to be created as are already needed for the number of people currently unemployed; and France, starting the decade with an unemployment figure of about 1 million, will have to at least double the number of jobs that were created

Table 18. Labour force by major occupational categories in France, the Federal Republic of Germany, the United Kingdom and the United States
(In thousands and *percentages of the total labour force*)

Category of worker	France 1962		France 1968		Federal Republic of Germany 1961		1970		United Kingdom 1966		1971		United States 1960		1977	
Clerical-managerial[1]	4009	20.2	5270	25.7	6065	22.1	7837	29.4	6542	26.3	8186	32.7	22007	31.5	41165	41.3
Production-labourers[2]	6795	34.3	7066	34.6	10337	38.5	9606	36.1	9781	39.3	10006	40.0	21951	31.4	32867	33.0
Agriculture-fishermen[3]	3940	19.9	3128	15.3	3626	13.5	2033	7.6	864	3.5	739	3.0	4261	6.1	2889	2.9

[1] Includes professional, technical, administrative, managerial, clerical and related workers. [2] Includes production workers, transport equipment operators, labourers and related workers. [3] Includes workers in agriculture, animal husbandry and forestry, fishermen, hunters and related workers.
Source: ILO: *Year Book of Labour Statistics, 1970, 1975 and 1978*, op. cit., table 2b.

Table 19. Net increases in the labour force 1950-1990, and unemployment 1969-79

Country	Net increase in the labour force,[1] in thousands		Annual percentage rate of growth of gross national Product per head[2]	Unemployment[3]					
	1980-90	1950-70		1969		1975		1979	
				'000	%	'000	%	'000	%
France	2 394	2 108	4.2	223	1.6	839	5.9	1 356	6.4
Federal Republic of Germany	1 883	3 600	3.4	179	0.9	1 074	4.7	1 171	5.1
United Kingdom	1 446	2 108	2.2	576	2.4	977	4.1	1 455	6.1
United States	11 290	21 848	2.3	2 831	3.5	7 830	8.5	6 431	6.4
Argentina	1 183	2 152	2.8	.	.	.	.	.	.
Bolivia	585	535	2.3	.	.	.	.	.	.
Brazil	13 274	12 253	4.8	.	.	.	.	.	.
Chile	950	713	0.9	.	.	.	.	.	.

[1] Source: ILO, *Labour force estimates and projections, 1950-2000*, op. cit., Vols III and IV, tables 1 and 6.   [2] Source: The World Bank, *World Development Report, 1978*, op. cit., pp. 76-77.   [3] Source: *Bulletin of Labour Statistics* (Geneva, ILO), 1st Quarter 1979. Strictly comparable figures for the developing countries in the table are not available.

in the past 20 years. The Federal Republic of Germany entered the 1980s with about 1 million people unemployed, and nearly 2 million jobs will need to be created during the decade to absorb the labour force increase.

In developing countries, in view of the general underutilisation of the labour force, the position is even worse. Table 20 gives orders of magnitude of underutilisation of the labour force in developing countries.

In Brazil the number of jobs needing to be created will be greater than the net increase in the labour force between 1950 and 1970; Bolivia and Chile are in the same situation. Only in Argentina will it be necessary to create fewer jobs than between 1950 and 1970.

The development of information technology affects precisely the sectors of the economy in which jobs have been created in the recent past. Agriculture has for long shown jobless growth; industry is also increasingly headed towards jobless growth—a trend which is being accelerated by automation. Now it is the turn of the service sector, which is expected to undergo a sharp slow-down in the creation of jobs owing to the direct and indirect effects of the technologies with which this book is concerned.

In the light of the trends described, predictions of 6 per cent unemployment rates in the 1980s in developed countries, and even 9 per cent for Western Europe by 1990, seem extremely conservative. As already indicated, these trends will deeply affect women and youth. An increase in the absolute number of unemployed above a certain level becomes a qualitatively different problem, especially when the unemployment is associated with geographical concentration, longer periods without work and a population unwilling to migrate in search of different employment.

In developing countries the problems will be even greater. According to a World Bank report, the developing countries will have to find jobs for about 550

Table 20. Preliminary estimates of underemployment rates in developing countries, by region, 1975

Region	Rural		Urban		Whole economy	
	A	B	A	B	A	B
Asia	40.8	39.5	24.9	23.2	37.9	36.4
Africa	43.7	41.0	28.2	25.1	40.8	37.9
Latin America	40.5	39.4	24.3	22.8	30.4	28.9

Key: A. Percentage of underemployed workers in the employed labour force.   B. Percentage of underemployed workers in the whole labour force, employed and unemployed.

Source: Peter Richards: "Poverty, unemployment and underemployment", in ILO: *Background papers*, Tripartite World Conference on Employment, Income Distribution and Social Progress and the International Division of Labour, Geneva, . . . 1976, Vol. I: *Basic needs and national employment strategies*, p. 8.

million workers from 1975 to the year 2000. According to table 15, the less developed regions generated 318 million jobs between 1950 and 1970. The figures in the table are not directly comparable to those of the World Bank because of different country classifications, but they serve as rough indications of the magnitude of the task.[16]

MEASURES PROPOSED FOR REDUCING UNEMPLOYMENT

A number of measures have been proposed to alleviate unemployment in the future; they can be roughly divided into four groups: reduced working time and work sharing; employment creation through new products and industries; incentives to small and medium-sized businesses; creation of employment in the service sector. These are not the only measures put forward: more radical and far-reaching changes have been suggested which imply different types of economic, social and political organisation, patterns of consumption and life-style. The four kinds of measures mentioned above are of a more traditional nature and are those currently being discussed by governments, management and trade unions in the industrialised countries.

1. Schemes centred on a reduction in working time propose a shorter working week, longer holidays, early retirement, and a different approach to overtime. Trade unions in Europe and the United States are committed, as a matter of policy, to a reduction in the working week from 40 to 35 hours.[17] In some places the shorter week is already a part of general labour agreements on wages, redundancies and productivity. Although a reduction of hours of work could reduce unemployment, it could also be outweighed by increased labour costs, the inflationary effects of which would damage long-term employment prospects and international competitiveness. A further difficulty is that the introduction of a shorter working week would probably be possible only under international arrangements, and even then the results in terms of job creation are uncertain, as several studies have shown.[18] The adoption of shorter working

hours by the industrialised countries would be an irreversible decision; through cultural channels or international bodies, it would be likely to influence work patterns in the industrial and service sectors of the developing countries, where the working week is generally longer than in the advanced countries of the West. How this will affect development prospects is completely unknown.

A more radical work-sharing scheme proposes transforming the rigid sequence of human life—education, work and retirement—into a flexible, recurrent system, in which it will be possible to alternate or combine education, work and retirement throughout a person's adult life. The advantage of this system is that it would allow maximum individual freedom while at the same time avoiding a sudden reduction in over-all working time.[19]

2. The creation of new products and perhaps new industries is another avenue being explored in the search for ways of alleviating unemployment. Many of today's commonly used products were impossible to imagine 30 years ago. This being said, it is also true that any future developments will be scientifically and technologically based as never before, so that they will usually be foreseen years in advance of their actual introduction as marketable products or processes: accurate predictions of the developments now taking place in micro-electronics were given by the industry ten years ago.[20] However, the timing of the development of some embryonic new industries, such as the space industry and underwater mining, is as yet a matter of speculation, whereas some established industries, including entertainment, travel and some person-to-person services, will undoubtedly be expanded in the fairly near future.

As shown in Chapter 4, most of the products and processes to which micro-electronics are applied are labour-saving. If the all-pervasive nature of the technology is taken into account in addition, the compensation theory which maintains that job displacement in one sector will be offset by employment creation in another begins to look distinctly unsound. In the electronics industry itself, the shrine of the new technology, the past and probable future trend is one of decreasing employment. Moreover, even maintenance and service activities in general and their supporting industries and suppliers will be firmly enmeshed in the micro-electronic web.

Increases in productivity could lead to the creation of some jobs over the medium and long term, provided that there is a growing market for the goods, that lower prices are effectively achieved, that capital is invested in productive, job-creating ventures rather than in labour-saving technology and, finally, that a high degree of labour force and resource mobility is achieved. These conditions have very complex implications, and are as unlikely to be fulfilled as the so-called "automatic adjustments" that are a feature of the compensation theory mentioned above. Historical trends in prices, market saturation of some lines, capital investment and resource mobility show that the possibility of meeting the assumptions of the compensation theory is remote, and that it would be extremely difficult at best.

3. Incentives to small and medium-sized businesses are based on the long-standing recognition of their economic importance and employment potential. "In most economies the numerical majority of a country's manufacturing enterprises (usually from 60 to 80 per cent in industrialised countries, and up to

or more than 90 per cent in developing countries) are classified as small. They may account for more than one-half of total employment, with the added possibility of being able to employ even more."[21] These companies are seen as a possible target group for employment creation, essentially on the basis of their likely capacity to absorb innovations and create new services. However, it has been mentioned before that the development of micro-electronics and the array of applications that accompany it are producing further concentration in many sectors of industry and the services. This only accentuates historical trends which have been well documented and which closely affect the survival of small businesses. In 1971 a study of OECD countries stated: "The general trend in all countries is towards structural reorganisation of industry leading to larger business units through such means as mergers and co-operation agreements, and this trend is naturally affecting the [small and medium-sized businesses]".[22] More recent evidence from the manufacturing and service sectors of Western Europe and the United States points in the same direction. In the United States a recent report by a special subcommittee of the House of Representatives on small businesses states: ". . . Economic concentration, riding the tide of an unprecedented merger wave, and monopoly power continue to gnaw away at the foundation of our competitive economic system—the small business."[23] The committee based its conclusions on the fact that from 1950 to 1976 the 200 largest manufacturers in the United States increased their share of market assets from 47.7 to 58.8 per cent of the total. In retailing, the share of large businesses in total sales increased from 29.6 per cent in 1948 to 44 per cent in 1972. In the Federal Republic of Germany a report by the small-business association argues that public action is necessary if further reductions in the market share of small businesses is to be avoided. The start-up cost for a small business rose from DM43,000 in 1963 to DM110,000 in 1973; it was thought that it might be as high as DM150,000 in 1979. Even in Japan, where big manufacturers are known for their traditional practice of subcontracting work to smaller firms, a reassessment of employment, working time and subcontracting policies is apparently regarded as inevitable, partly as a result of automation.[24]

4. Employment creation in the service sector is another possible remedy for the displacement of labour by the new technology. The underlying assumption is that services have, in the past few decades, always played a labour absorption role. There are three main reasons why employment could expand in the service sector:

   (a) the need to maintain and develop services such as health, education and social services; demands on the educational system are particularly strong owing to changing job patterns and mobility;

   (b) the emergence of new needs and demands, on the assumption that increasing income and social complexity lead to higher consumption of services; and

   (c) the fact that even in an open economy, services in general are protected from international competition.

Many different activities are performed under the umbrella of services. It is the labour-intensive or person-to-person type of services that will probably be most extensively developed in the future. However, two additional points need to be borne in mind: the first is that the greatest employment loss seems likely to take place in the office sector, as has frequently been shown in this book. The second is that a shift is taking place from services to goods, as durable consumer goods are being substituted for labour-intensive services. In *The coming of post-industrial society*, published in 1973, Daniel Bell argued that in the industrially advanced countries the economy would be increasingly centred on the provision of services.[25] More recently Jonathan Gershuny has demonstrated that as incomes have increased the general trend in the United Kingdom, as revealed by statistics of household expenditure from 1954 to 1974, has been towards the acquisition of household goods as substitutes for the purchase of services. Examples of this are the substitution of individually owned cars for public transport, washing machines for laundries and television and radio for cinemas and live opera and concert performances. He concludes:

> . . . Consumption of goods represents a fundamental change in the nature of economic activity. Instead of capital investment taking place in industry, and industry providing services for individuals and households, increasingly, capital investment takes place in households, leaving industry engaged in what is essentially intermediate production, making the capital goods—the cookers, freezers, televisions, motor cars—used in home production of the final product. This is the trend towards the do-it-yourself economy—almost the antithesis of Bell's service economy.[26]

Gershuny further substantiates his point by showing that, despite the fact that about half of the working population in the United Kingdom is employed in the tertiary sector, less than a quarter is engaged in the provision of services in the sense of person-to-person activities. He explains that the growth of service employment is a result of the extension of the production of material goods. In short, there is reason to believe that the expansion of employment in the service sector will not be as rapid as was expected in the later 1970s. It seems that the growth of service industries is closely linked to the use of service machines as opposed to an increase in the use of labour. Current technological changes will further this process both within the home and outside it through the development of cheaper, more reliable household appliances to perform an increasing variety of functions. This is not to say that services will not be created or expanded, but again the point is where jobs will be created and how many.

Changes in the service sector can be attributed to a number of factors ranging from an increase in household incomes and expenditure to a shift in popular demand towards more individualised, made-to-order products that cannot be adequately provided through mass production. What we appear to be witnessing now is a convergence of technology with a search for individuality.

Services such as education and health are starting to be "materialised". (In the industrialised market economy countries the prospects of increasing employment in the educational field already seem poor in any event in view of the expected decline in the size of the school-age population in the 1980s and 1990s. Many countries have already cut down their teacher-training programmes.)

In the short term there will be a great need for specialised personnel, particularly programmers and electronic engineers. The technical obsolescence of recently acquired knowledge, which is a real problem in the electronics industry, will call for the constant retraining and recycling of staff.

Parts of the service sector can, however, afford lower levels of productivity since they are not directly confronted with outside competition. This could very well make it possible to slow down the pace of technological applications, since it is not the main focus of investment attention and policy can very often be directed by central authorities.

The changing pattern of individual and collective behaviour must be considered. A mechanical extrapolation of technological trends does little to help understanding of the issue. There are a number of cultural trends and traditional attitudes that impede or slow down technological change; at the same time, technology creates social and attitudinal changes. The relative failure of fast-food restaurants in France, or the persistance of nepotism as opposed to scientific management in family businesses, could be mentioned as examples of the influence of traditional attitudes. Home delivery of milk and its social role are also related to this issue. There are a number of other traditional services that are not suitable for automation owing to the absence of sufficient economies of scale; these include house painting, cleaning, shoe repairing, and the retail marketing of fresh food. Although automation is being applied to some equipment in these fields such as self-programmed vacuum cleaners or programmed robots for office cleaning, they are only in the prototype stage. Similarly, bars and coffee shops could be automated but they would lose one of their main functions, which is to provide opportunities of meeting people. Where the new technology is being applied in the service sector, the resulting decrease in person-to-person services is producing a cultural reaction which could be of great importance in the future. An example of this is the growing enthusiasm for "natural" forms of medicine and an emphasis on the healing power of human contact. Another similar example is the reaction to the chemical content of foods together with doubts about their nutritional value, which have further increased the demand for high-quality ("organic", for lack of a better word) protein-rich foods which are labour-intensive in production and distribution. At the present day these trends are not dominant, but as incomes increase and the quality of life becomes more important, they could affect employment.

Furthermore, the response to centralisation and developments in the cultural field could create employment, as could the increasing awareness of ecological and environmental issues. Higher levels of unemployment, more leisure time and early retirement could create the need for social services to deal with possible increases in youth delinquency and divorce rates and, more importantly, to maintain the mental health of the population in countries in which it still relies to some extent on the work ethic as its standard of self-value. In addition, high unemployment, over longer periods, could cause the development of an informal or marginal sector outside the tax economy and beyond legal regulation. Institutional dynamism, or the survival instinct, will also have an effect on levels of employment in the service sector. Cynics suggest that the smaller a country's fleet, for example, the bigger the admiralty staff is likely to be. The implementation of new government regulations also creates employment, although there is no reason why government activities could not be largely automated.

THE GENERAL PROSPECT

The combination of factors mentioned in this section reinforces the need for industrialised market economy countries to pursue a policy of readjustment. In the process they will continue to set the standard of competition and the pace of innovations. The provision of employment, particularly the number of new jobs to be created and their location, will remain subject to much debate. It is safe to say, however, that no potential means of job creation is free from substantial doubts about its likely performance. Nevertheless, it is also clear that the maintenance of international competitiveness, through a combination of the elements mentioned at the beginning of this chapter, will to some extent neutralise labour displacement in some sectors. However, maintaining competitiveness also results indirectly in labour displacement through the process of business concentration. As a general rule, it could be said that the lower the prices, particularly of micro-electronics, the harder it will be for companies that are not vertically integrated to compete.

The long-term trends discussed above seem to suggest that we may be facing a transition towards a society that no longer needs to make full-time use of its entire potential labour force, at any rate under current employment conditions. In view of this possibility, the arguments in favour of different life styles and industrial systems are more persuasive, and will become increasingly so as the micro-electronics revolution progresses. A transition of this nature in industrialised market economy countries will not be free of turmoil. There are historical precedents for transitions involving entire sectors. Agricultural improvements, including mechanisation and standardisation, led in the long run to an absolute decrease in the rural labour force of most countries. This took place despite the fact that it was accompanied by a reduction in working time throughout society. Industry and services took over as job creators. The decline of job creation in industry has been evident for some time and the tendency is towards a reduction in absolute numbers as well. Services have remained the universal job provider, but can no longer be relied upon in this respect because of the introduction of micro-electronic technology. The effect on employment must be assessed in the light of a well documented historical trend that is related not to micro-electronics as such but, more generally, to mechanisation, automation and changes in products.

Notes

[1] The World Bank (Washington): *World development report, 1978*, p. 9.
[2] ACARD, *The applications of semiconductor technology*, op. cit., p. 8.
[3] Nora and Minc, *L'informatisation de la société*, op. cit., p. 40 (ILO translation).
[4] McLean and Rush, *The impact of microelectronics on the UK*, op. cit., p. 51.
[5] ibid., p. 20.
[6] ibid., p. 27.
[7] ibid., p. 38.
[8] ibid., p. 49.
[9] Nora and Minc, op. cit., p. 43 (ILO translation).

[10] For examples of material issued by different unions and trade unionists on micro-electronics, see Jenkins and Sherman, *The collapse of work*, op. cit.; APEX, *Office technology*, op. cit.; Trades Union Congress: *Employment and technology*, . . . TUC interim report . . . prepared by the General Council for consideration at the TUC consultative conference . . . 1979 (London, 1979). See also European Trade Union Institute (ETUI): *The impact of microelectronics on employment in Western Europe in the 1980's* (Brussels, 1979).

[11] Freeman, *Government policies for industrial innovation*, op. cit.

[12] *Prognos Euro-report 1979: Annual report of the economic and demographic trends in Western Europe and the United States* (Basle, 1978).

[13] The argument that the effects of technical change should be assessed in the light of the transition from agriculture to industry and services was developed by Professor C. Freeman in the UNESCO-Bariloche workshop on analytic techniques for long-term development issues, held at the Institute of Development Studies of the University of Sussex from 20 November to 2 December 1978. It was subsequently suggested to me by Mike Hopkins (International Labour Office).

[14] For details of this shift in selected developed and developing countries, from the beginning of the century up to around 1970, see Yves Sabolo: *The service industries* (Geneva, ILO, 1975), tables 2 and 3.

[15] *OECD Observer*, No. 90, Jan. 1978. For a detailed treatment of this point, see Peter Melvyn and David H. Freedman: "Youth unemployment: A worsening situation", in Freedman (ed.): *Employment outlook and insights*, A collection of essays on industrialised market-economy countries (Geneva, ILO, 1979); see also ILO: *Young people and work*, Report II, Third European Regional Conference, Geneva, . . . 1979.

[16] The World Bank: *World development report, 1979*, p. 110 and Chapter 4.

[17] *Social and Labour Bulletin* (Geneva, ILO), Sep. 1978.

[18] "Measures to alleviate unemployment in the medium term: Work-sharing", in *Department of Employment Gazette* (London), Apr. 1978, pp. 400-402. This subject is still under debate and evidence is contradictory with regard to potential job creation. The area which might offer the biggest potential is the elimination of overtime, but this could have a tremendous effect on incomes. The statutory provisions available to enforce such a measure, and even its practicality, are not clear. For different reports on the state of the debate, see the *Financial Times*, 2, 22, 23 and 27 June 1978.

[19] Louis Emmerij: *The social economy of today's employment problem in the industrialised countries*, IEA conference on unemployment in Western countries, Strasbourg, . . . 1978.

[20] Solomon Miller: "Technological trends in the component industries", Annex I of OECD: *Gaps in technology: Electronic components* (Paris, 1968), p. 127.

[21] Philip A. Neck: "Role and importance of small enterprises", in ILO: *Small enterprise development: Policies and programmes*, Management Development Series, No. 14 (Geneva, 1977), p. 10.

[22] OECD: *Problems and policies relating to small and medium-sized businesses*, Analytic report drawn up by the Industry Committee of OECD (Paris, 1971), p. 49.

[23] United States Congress, House of Representatives Small Business Subcommittee (Washington, 1979).

[24] For an important discussion on this point, see University of Sussex, Science Policy Research Unit: *Small and medium-sized manufacturing firms: Their role and problems in innovation: Government policy in Europe, the USA, Canada, Japan and Israel*, Report prepared for the Six Countries Programme on Government Policies towards Technological Innovation in Industry, by R. Rothwell and W. Zegveld (1978, 2 vols.). See also *Financial Times*, 4 Dec. 1978 and 16 Feb. 1979.

[25] Daniel Bell: *The coming of post-industrial society*, A venture in social forecasting (New York, Basic Books, 1973).

[26] Gershuny, *After industrial society*, op. cit., p. 81.

EFFECTS ON DEVELOPING COUNTRIES 7

Technological change conditions, but does not determine by itself, the distribution of manufacturing facilities, which is one of the most visible facets of the international division of labour. There are many other relevant factors such as currency values, trade arrangements, government policies, foreign exchange regulations and peculiarities of markets, to mention a few. Although the importance of the above-mentioned factors is fully recognised, the intention in this chapter is to emphasise changes caused by information technology in developing countries.

RELATIVE PRODUCTION COSTS

The discussion in previous chapters on the effects of information technology has suggested that automation is becoming competitive with cheap labour, at least in some sectors. There are at least five main reasons for this:

(1) By decreasing the proportion of direct labour costs in total business cost, automation erodes the advantage of low labour cost and makes possible the manufacture of formerly labour-intensive goods in developed countries.

(2) The application of the technology to organisation reinforces the management, marketing and co-ordination superiority of developed countries, by increasing the efficiency and lowering the cost of recording, processing and retrieving information.

(3) In order to improve products and processes, management in some traditional mass production industries is beginning to promote research and development, and the industries themselves become highly capital-intensive when the improvements are introduced. The radical improvement of older products and the introduction of new ones require more sophisticated and integrated manufacturing processes. In this way, more and more industries cross the barrier into the high-technology category where design and quality become essential to the acceptability of products.

(4) The technology encourages further industrial and service concentration and vertical (forward) integration, which are accompanied by streamlining in

certain sectors and increased efficiency in the use of resources. A different result of this process, however, can be the creation of new opportunities for newcomers who take advantage of the loss of production flexibility and innovative capacity in established firms.

(5) The use of micro-electronic technology produces a saving in capital per unit of output, but an increase per worker employed and, thus, of capital intensity. Formerly labour-intensive activities are thus becoming capital-intensive, with important consequences for capital requirements and formation.

In order to illustrate the above-mentioned trends, the situation in two broad industrial sectors will be briefly analysed—textiles and clothing, and electronics. It will be seen that in those two sectors micro-electronics will have a significant effect by eroding the advantage of cheap labour. Much research will be needed to fully understand the long-term implications and side effects and, beyond that, the consequences for industrialisation policy as a whole. Improvement in manufacturing productivity is only part of a general improvement on all fronts. This progress is based on technologies and practices which are far better developed in the industrialised market economy countries than in the developing countries. In addition, the introduction of micro-electronics reduces the possibilities of establishing or locating manufacturing facilities in developing countries. As traditional industries are transformed by changes in products or processes, they will move out of the reach of developing countries.

Textile and clothing manufacture

In 1976, 21 developed countries imported manufactured goods to the value of $27,000 million from developing countries. Of this total textiles, clothing and leather and footwear represented 11.4, 23 and 5.8 per cent respectively.[1]

The textile and clothing industry embraces an enormous variety of sectors and subsectors, in which different materials and processes are used and different goods produced, and in which market conditions vary. Some of these sectors, such as that of man-made fibres, are among the most highly automated and capital-intensive in industry, while others remain labour-intensive and fragmented. Some of the differences can be attributed to cultural differences and other factors peculiar to individual countries. This somewhat daunting picture of the diversity of the industry is given in order to underline the difficulty of assessing the effect of technological change upon it: generalisations indicating main trends may not be applicable in all cases.

Some developing countries have put considerable reliance on the textile industry as a means of promoting industrial development because of its heavy demand for labour, its use of local raw materials and traditional techniques, and its low investment-output ratio. The standard pattern has been import substitution, supply of the domestic market and, in some cases, exports. In these countries the industry remains largely labour-intensive and it has been noted that even when advanced technology is employed, productivity is significantly lower than in the developed countries.[2]

In developed countries too there are some sectors which remain fragmented and labour-intensive. This lag in certain sectors has been due to the complexity of production processes and to constantly changing design which does not justify purpose-built equipment. However, the general trend has been and is towards automation and capital intensiveness. This trend in the developed countries will inevitably have repercussions in the developing countries. The use of self-programming robotic arms for cutting, and of computerised systems for design, producing patterns, monitoring the quality of fabrics and guiding laser beam cutters, is changing the face of the industry. Microprocessors are being used to control sewing patterns, fast stitching, knitting heads (instead of the punched cards associated with the Jacquard loom), and ink injectors that can be rapidly adjusted to produce different designs and colours. These are part of a growing number of applications aimed at providing "a total system", which means the use of computerised techniques to detect flaws, keep track of patterns and orders, monitor the progress of work throughout the plant and automate the matching of patterns and the cutting and sewing. These applications save labour, skills and materials. In materials, for example, the saving ranges from 8 to 15 per cent. The techonological changes are not only related to micro-electronics but include the introduction of new high-capacity machinery and continuous manufacturing processes.[3]

In the short run automation will be concentrated in the manufacture of high-quality products, particularly fashion products. In this part of the industry, developed countries have a lead time of six months or more, operate close to the fashion centres and have been heavily protected. The impetus towards automation mainly derives from fierce competition between developed countries. Developing countries have not competed in this sector in the past, nor is it likely that they will do so in the future.[4] The combined effects of automation and of innovations based on electronics are reinforcing the transition of the industry as a whole towards the high-technology category, relying to an increasing extent on research and development and the use of computer software. There will be improvements in design of products and processes, in quality, and in planning and marketing, in all of which respects developed countries already enjoy traditional advantages. In fact, the links between end-producers, machine manufacturers and electronic or computer firms have been multiplied in a joint effort to achieve higher levels of automation.

The readjustment policies of developed countries have greatly encouraged the automation and modernisation of the industry. Automation is seen as the only way of maintaining the industry in the industrialised countries of the Northern hemisphere, albeit with reduced employment. At the same time this process is leading to further industrial concentration through mergers and take-overs and, inevitably, to an uneven distribution of the technology.[5] Looking at these trends, a report by the ILO states: ". . . The competitiveness of low-labour-cost firms with labour-intensive techniques is continually being eroded by the installation of high-productivity capital-intensive machines in developed and some developing countries".[6]

Forecasts suggest that by the year 2000 textile production will be so capital-intensive that labour cost differentials will no longer play the decisive role in total product cost that they do today.[7] This forecast was made in 1976, before the large-scale introduction of micro-electronics, and all the evidence suggests that the change is far more rapid than was then expected. Nevertheless, erosion of labour cost advantages will take much longer in some developing countries than in others. One study of 20 countries shows that in textiles, the lowest labour cost (Pakistan) was one-thirtieth of the highest (Belgium).[8] However, it should be added that in this industry proximity to the main markets and end users is a considerable advantage because of the incidence of transport cost on the final price. Protectionist measures, particularly quotas, have forced exporters from some developing countries to move up market in order to obtain more value for the same quantity of products, thus forcing their own industry to upgrade and enter a market where they have the least advantage. The countries and territories that can upgrade successfully will tend to monopolise developing countries' exports by decreasing the competitiveness of others. It is already becoming evident that, in the upper part of the textile and clothing industry and also in new products, competitive edge is now provided by technology and not by labour cost. Over the medium term, this will become true of the industry as a whole.

Given the state of research on these problems, it is not possible to provide a precise time scale or a quantification of side effects. What is clear, however, is that this industry will not enjoy the same importance for developing countries as in the past, nor will it be a low-investment short cut to employment provision and foreign trade earnings. Once technology gives the competitive edge in the industry, it is only a question of time before relocation of plants begins.

The electronics industry

The electronics industry puts out a wide range of products, including components, computers, instruments and consumer goods; special attention will be given here to the production of components, particularly integrated circuits.

Since the 1960s the semiconductor industry has been searching for low-cost labour for the testing and assembly of chips, this being the most labour-intensive phase of their manufacture. Some developing countries have adapted their foreign investment, tariff, tax and labour policies in order to facilitate the establishment and operation of high-volume industries producing goods with low added value.[9] Although the establishment of production plants by foreign firms in circumstances such as these gives rise to certain economic distortions, such plants have been an important source of foreign exchange, and to a lesser extent of employment, for the host country. However, a number of factors have recently converged which cast a shadow over the medium-term prospects for export-oriented electronics industries in developing countries. Automation of the former highly labour-intensive assembly and testing of chips is about to begin; the equipment has already been developed. International Telephone and Telegraph (ITT), which is one of the pioneers in this field, sees automation as the only way of reducing costs to a minimum, increasing efficiency and maintaining manning levels in developed countries, in this particular case the United Kingdom; in addition, product manufacture will be close to the majority of end-users. This trend has been present in the industry for some time. In 1977 W. G. Oldham stated:

> The traditional cost-saving technique has been to employ less expensive overseas labour for the labour-intensive packaging operation. As the cost of overseas labour rises

and improved packaging technology becomes available, overseas labour is gradually being supplanted by highly automated domestic assembly.[10]

The Group General Manager of ITT Semiconductors puts the argument in the following way:

A typical semiconductor company with a turnover of $200 [million]/year must expect today an annual price decrease factor of approximately $30 [million], necessitating corresponding manufacturing cost reductions of $40 [million], even taking into account today's world of inflation, increasing labour and material costs. These cost reductions must result from increased mechanisation, higher yields and increased volume.

The relative advantages of increased mechanisation versus offshore assembly are reviewed next and the conclusion comes down firmly in favour of the former, permitting closer customer contact, elimination of logistics problems, and upgrading of local or national technology, without which complete business transfer to low labour cost areas will occur (e.g., the US TV industry, and the European camera and textiles industries).

The conclusion is that the successful European semiconductor manufacturer must have the complete production cycle in one manufacturing location, worldwide marketing, fast market introduction and close cooperation with end user industries, e.g., computer, entertainment, telecommunications and automotive.[11]

This point of view is substantiated by the fact that none of the firms engaged in the latest ventures in the field of integrated circuits (Inmos, GEC-Fairchild, ITT and National Semiconductors in the United Kingdom, Thomson-Motorola in France and Mostek in Ireland) contemplate the establishment of offshore installations in low-wage countries. The transition to automation and geographical concentration is also a result of generous incentives and heavy investment by some developed countries, where governments wish to create jobs and develop national capabilities in a key sector.[12] Protectionist measures, currency values and transport costs for finished goods have meant that Japanese firms, for instance, are investing in manufacturing plants in the United States and Europe. This will widen the national bases of the industry and the possibilities of protecting a firm established within a foreign country against any possible influx of even cheaper goods.[13] The tempo at which offshore installations are phased out will be temporarily slowed down owing to persistent shortages of some components. This shortage, which is due to unexpectedly strong demand mainly from the automobile and computer industries, will not alter the general trend in the medium term.

Some operations, however, will continue to be performed in developing countries in order to exploit national and regional markets, where there are local incentives, including in particular the likelihood of receiving contracts for locally assembled equipment. They will stay for the testing and assembly of simpler chips for which the production volume does not fully justify automated plants. There is also a tendency to transfer offshore installations from higher-cost Asian countries such as Hong Kong, the Republic of Korea and Singapore to countries such as Thailand and the Philippines in an effort to cut costs even further. Governments could be instrumental in retaining manufacturing facilities if they follow policies encouraging vertical integration of the industry. This is the case in the Republic of Korea, where it is hoped to be producing a complete computer system by 1986. In order to do this, the Republic needs to maintain and develop the production of the most advanced integrated circuits. It is doubtful whether this is a wise and profitable policy in view of current trends in the highly industrialised countries elsewhere. Brazil has opted for a different strategy, reserving sections of the market for domestic assembly and manufacture in an attempt to increase the national integration of production. Semiconductor firms which test and assemble chips will continue to operate in order to profit from the potential growth stimulated by preferential treatment for locally based manufacturers (Brazil has reserved the microcomputer and minicomputer section of the market for itself, but this policy could be altered in the future). Both of these countries remain technologically dependent on multinational firms, and it is unlikely that these firms will transfer technologies that will allow competition with their own factories by others operating abroad.

Micro-electronics have rendered obsolete the traditional method of measuring the degree of national integration of a given industry. This method was useful for the purpose of monitoring transfer of technology and creating incentives for development of a national industry. Under the current system a combined weight-volume-value index is used to assess the extent of national integration of a product. Integrated circuits embody low weight and value but high technology, which gives a misleading picture of the extent of national integration: in fact, the worth of an electronic product lies in its use of integrated circuits and other electronic components, although they may have only a very small weight, volume and value themselves. Thus developing countries would appear to doing better than is actually the case.

The developing countries with an export-oriented industry have promoted electronics in an effort to diversify towards high-quality finished products. This has rendered the industry more vulnerable to current changes since the phasing-out of offshore installations in components will take away significant business. A case in point is the Republic of Korea, in which the electronics industry has enjoyed spectacular growth. In 1972 the country's exports of electronic goods were worth $177 million and its imports $142 million. By 1978 exports were up to $1,396 million and imports up to $1,000 million; of the total exports, components represented $614 million or 44 per cent (integrated circuits 27 per cent, semiconductor devices 6 per cent and other parts 11 per cent).[14] Already by 1973 the domestic industry could not meet the demand for components, which increased very rapidly as the industry diversified. In-depth technological developments can probably take place in some marginal sectors, but not in the electronics industry as a whole because of the huge current and future cost of technological innovation. Export-oriented electronic industries in several other countries are in a similar vulnerable situation.

The rate of plant relocation will depend on the investment policy and amortisation of each company. It is clear, however, that it will be rapid: ITT expects to become the third-largest or fourth-largest producer with its new automated plants in developed countries, thus compelling other manufacturers to follow suit.

Developing countries cannot develop an information industry able to compete on the international market. Some countries such as China and India have developed significant indigenous industries and are moving towards large-scale integration in the component field. This process is accompanied by strong protectionist measures and by the application of technical obsolescence criteria different from those prevailing in the international market. The cost of the products, however, is considerably higher. The obsolescence of a given product is not intrinsic to it but related to the new products available. For the purpose of building indigenous technical capabilities, developing countries need to maintain criteria of obsolescence different from those of the highly industrialised market economy countries and a policy of highly centralised purchasing and application of computer technology.[15]

The possibility of combining low labour costs with high technology is doubtful, except perhaps where highly sophisticated machine tools are used. In that case parts assembly, which may not warrant investment in automation but which may make profitable use of cheap labour, can be substituted for precision engineering. There may be possibilities of this kind in the watch and clock industry. However, in most cases product changes lead to industrial concentration and economies of scale rather than deconcentration. A more general obstacle to the solution of combining low labour costs with high technology is that the technologies are controlled by a few multinational companies with world-wide production and marketing facilities. In the field of integrated circuits, it is their explicit policy to prevent the transfer of technology.

To turn to another aspect of the new technology: micro-electronics is software-based and requires a set of skills which are more sophisticated and difficult to acquire than traditional skills of a mechanical nature. Particularly important in this respect is the shift towards the design of entire systems of production instead of separate items of machinery; this calls for systems engineering skills which are acquired only after high-level training and long experience. The distinction between mechanical and electronic engineers is tending to disappear in highly industrialised countries, whereas developing countries are reproducing this now obsolete division of skills. Even in developed countries, the transition to electronics in formerly mechanical industries such as watches and office equipment has shown that the retraining of personnel is almost impossible and involves high cost and risk, especially when dealing with older workers.[16]

The efficient use of software, besides being a highly sophisticated skill, depends on experience gained in solving programming problems. Thus the increasing development of software contributes further to unequal development between industrial market economy countries and less developed countries. There may be a few exceptions to this general rule in some developing countries where the conversion rather than creation of software may provide some employment.

INFORMATION POWER

The setting up of supranational networks for the handling of information is another important aspect of current technological changes. The question of transborder data flows would deserve a chapter to itself because of its complexity and far-reaching consequences. Two points will be mentioned here, one related to research and development and the other to a more general economic and political issue.

Computer technology has been widely used for scientific and technological research. Information technology has opened up new horizons through its veritable cornucopia of advantages which, by this stage of the book, are thoroughly familiar—economy, increased capacity and reliability, improved storage and retrieval, incorporation of telecommunications. Data banks and bases have served scientific research since their origin in the late 1950s, and on-line information searches are growing at a rate of 30 per cent a year. It is not difficult to locate the most important users and owners of the information stored in data banks. Of the 2 million computer searches carried out annually, 75 per cent originate in the United States. The individual firms holding most of the world's data bases are Lockheed and System Development Corporation, which account for 75 per cent of the European market and 60 per cent of the United States market. In 1977 Lockheed had about 100 of the 500 publicly available data bases in the world on its computers.[17]

The availability of data will increase tremendously in the future as costs are reduced and new systems are set up. One of these new systems is Euronet, which is being set up by the European Communities and is initially aimed at the storage and retrieval of scientific and technical information. It will be run by a consortium of European post office departments and is scheduled to become operational in 1979.[18] Such networks allow scientists in different places to participate in an experiment taking place at a single geographical point. Data gathered in country A can be analysed in country B with the help of additional information from country C. The countries can thus help each other to solve problems, or add to existing knowledge.

At the same time, the increasing cost of research and development has encouraged co-operative endeavour in a number of fields (basic research, aerospace, defence, nuclear energy and multinational joint agreements in, for example, the automobile, electronic and chemical industries). The convergence of high technology resources is an important element in the policy followed by the highly industrialised market economy countries, and is greatly encouraged by competition among developed countries and between multinational companies. These arrangements increase research and development capacity, with the help of information technology.

More often than not, "intelligent equipment" is not only labour-saving but also skill-saving. Since developing countries do not possess a large skilled labour force, their shortage of skills can be overcome by the introduction of intelligent equipment. This leads to a circular situation in which skills are not developed because they are not needed, and the capacity for indigenous research and development is undermined.

These developments will change the pattern of technology transfer and will further increase the technological and scientific gap between developed and developing countries. Indeed, the decreasing cost of communications and data transmission and the concentration of information in developed countries means that it is becoming cheaper for institutions and firms in developing countries to have their problems solved for them elsewhere than to develop local research facilities. J. M. Carroll has identified a further dimension of the process: "Data which provides the basis for decision-taking will flow toward richer, more developed nations; information reflected in decisions already taken will flow toward poorer, less developed nations."[19]

Information stored in remote locations may not only be of a routine nature;

it could be or become vital to a country's socio-political, as well as economic, functioning. The Canadian Minister of State for Science and Technology, opening a congress of the International Federation of Information Processing in 1977, stated the consequences:

... The problem of transnational data flows has created the potential of growing dependence, rather than interdependence, and with it the danger of loss of legitimate access to vital information and the danger that industrial and social development will largely be governed by the decisions of interest groups residing in another country.

Here we are dealing with something intangible: you can't pick it up, you can't count it as it crosses an international boundary, and most of it relates to intracorporate requirements, rather than to something for sale in the market place. We may not be able to rely on traditional means of carrying out—or measuring—international trade in these invisible intangibles. There is a growing need for international agreement on a generally acceptable set of ground rules for dealing with these perplexing problems. ...

We are all "developing" nations in the new information society, and solutions to these problems cannot always be found in tradition or precedent.[20]

OTHER EFFECTS

Three other effects on developing countries will be briefly mentioned. First, low labour costs will become increasingly advantageous for such purposes as the conversion of software, particularly in countries with large numbers of well educated workers. Work on software remains highly labour-intensive despite automation and standardisation, and a number of European-based software companies are transferring their conversion work under contract to countries such as India. Coupled with economic advantage is the added credibility of the final product, since it is thus tested in conditions that can be commonly found in other developing areas.[21] The employment effect of the transfer of software conversion work to low-wage countries is difficult to assess at this stage, but it is likely to be relatively insignificant and localised, and will not absorb large numbers of workers. The effect might be greater if it became the general practice to use well educated workers for this task. It must be emphasised that the research and development which produce the essentials of the software are carried out in the developed countries, and this state of affairs will not change: because software is to an increasing extent the marginal factor that gives the competitive edge, its transfer is subject to strict controls.

Secondly, developed countries are likely to take a tougher attitude towards developing countries. This tougher attitude may find expression in intensified national readjustment policies, protectionism and the promotion of traditional domestic industries. Some of the measures taken to cope with national concern in the developed countries will have global effects. Although work-sharing schemes and the 35-hour week have not received support in many countries, it is possible that some arrangement of this nature will be introduced in the future. A change in working hours will follow in developing countries in much the same way as, for example, the employment of women workers and a concern with environmental issues. The precise nature of the effects is difficult to guess at the moment; more information on this topic is needed.

Thirdly, current changes start a train of others. A case in point is the introduction of glass fibre for use in telecommunications (fibre optics) as a substitute for one of the most important applications of copper. Although full commercial development is not expected until the mid 1980s, the fact that the technology will be available conditions all planning concerned with telecommunications. Since, in addition, graphite and resins are being substituted for other metals, solid-state technology is superseding electrical, electromechanical and other components and products will be redesigned, it is clear that the demand for some raw materials will be altered.[22] Changes of this nature will have an enormous effect on some developing countries and technological forecasting in these areas is becoming essential for investment planning and long-term development prospects.

EMPLOYMENT AND INCOME DISTRIBUTION

Most of the evidence accumulated in respect of employment and income distribution refers to direct use of computers, particularly for data processing in office work; however, this affects a very small proportion of the total labour force and can serve only as a general indicator of possible developments.

In the United States, which is the most computerised country in the world, data processing personnel represented almost 1 per cent of the total labour force in 1974 and is expected to increase to just over 1 per cent by 1985—an over-all increase in absolute numbers of 29.4 per cent (853,000 in 1974 to 1,104,000 in 1985). This is in the context of a total growth of the labour force of 20.3 per cent for the same period (85.9 million to 103 million).[23] In India, on the other hand, it has been estimated that employment in computer installations does not amount to more than 0.4 per cent of total employment in organisations having computers,[24] and hence a tiny proportion of the total labour force. It should be borne in mind that computer utilisation is just the tip of the iceberg with regard to the use of information technology, but that it can serve as an indicator with regard to the effect of technological change on employment.

Both the short-term and the long-term effects are different in developing countries from those in developed countries owing to differences in the structure of industry and services. The development of information technology creates jobs at the manufacturing and end-user level by creating new services and new types of work in, for example, data processing departments in many sectors of the economy. Much of the existing evidence refers to the effects of applications of the technology to data processing, and some of the conclusions to be drawn are contradictory.

Two rather old studies, one in a developed country (United Kingdom, 1965), the other in a developing country (India, 1972) conclude that computerisation leads to a net loss of jobs or to a loss in potential employment in functions related to data processing.[25] Experience in the United Kingdom showed a 10 per cent fall in the number of jobs at a constant level of business and a 12.5 per cent loss of potential employment for an increased volume of business. In the case of India the study showed that computerisation led to decreases in employment, but that this was offset by jobs created in data processing. The over-all marginal increase in employment came from new applications, and there was a bonus in increased flexibility. The same survey demonstrated that there was a considerable loss in

Table 21. Employment effects of computerisation in the United Kingdom and India

Effect	United Kingdom	India
Total loss of jobs as a result of data processing	− 14.5	− 4.9
Additional posts created	+ 4.5	+ 7.4
Posts which would have arisen but for automatic data processing	− 12.5	− 19.1
Net change in total employment	− 22.5	− 17.6
Net reduction in number of persons per average organisation[1]	*90*	*112*

The average organisation using automatic data processing is larger in India.

Sources: United Kingdom, Ministry of Labour, Manpower Research Unit: *Computers in offices*, Manpower Studies, No. 4 (London, HM Stationery Office, 1965), p. 23, and Government of India, Ministry of Labour and Rehabilitation: *Report of the Committee on Automation* (Chairman: Prof. V. M. Dandekar) (1972), as cited in R. A. Kanga: *Computer policies for developing countries* (manuscript, 1977).

potential employment. Although the Indian sample covered only three organisations by comparison with 562 in the United Kingdom, and although the environment in industry and services was very different in the two countries, the net reduction in total employment was comparable. Table 21 summarises the results in both cases. It is very difficult to assess the findings of the two studies. The number of job losses may be underestimated and there may well be workers who are underemployed but maintained on the payroll. Moreover no consideration is given to possible employment effects in other sectors of the economy as a result of increases in productivity and because the use of computers can remove a number of bottlenecks in industry, services and government. Both studies neglect the short-term increase in employment in related industries such as paper manufacturing, components, peripherals and air conditioning. In fact, a subsequent study in India shows that, when these factors are considered, there is a net gain in employment. Table 22 gives a summary of these findings.

Before commenting on these findings, it is necessary to mention another piece of research carried out by the ILO at the beginning of the 1970s. The general conclusion of that study, which covered Bangladesh, Brazil, Ethiopia and India, was that "there can be no doubt about it: a slowing down in employment growth . . . was the most common economic effect."[26] All the studies mentioned agree on the fact that there is a loss of potential employment; this will have special effects in developing countries with their high underemployment rates and growing educated unemployment. At the same time, the job creation potential of computerisation, argued in some of the studies, needs to be qualified on at least two grounds: first, the employment created in manufacturing, assembling, peripherals and related industries is valid only in very few developing countries, since in most of them equipment and ancillary goods are imported. Secondly, at the time of these studies the current tremendous development of micro-electronics had not begun. In addition, the dynamic of

Table 22. Net increase in the number of jobs as a result of the use of computers in India

Computer installations	9 000
Manufacturing	1 519
Sales	1 151
Ancillary units	1 000
Total employment created	12 670
Employment displaced	− 8 000
Net increase	4 670
Net increase per computer	*18.4*

Source: Kamta Prasad and Promod Verma: *Impact of computers on employment* (Delhi, Macmillan, 1976), p. 125.

the technology has altered the personnel requirements at several levels: for instance, whereas much of the job creation in the past has been for key-punch and verifier operators, their functions are now being taken over by direct input techniques. The staff cost of machine operation remains one of the most expensive aspects of data and word processing, and the industry has concentrated on simplifying inputs, for example by using direct input, optical readers and computer input microfilm. Many of these input techniques remain expensive and rather unusual, but there has already been a net loss of jobs for key-punch operators in data processing. The estimates of the Indian study, which shows a net gain in employment, were based on an average employment gain of 18.4 jobs per computer. When employment among data processing staff is examined by job function, it is seen that key-punch and verifier operators represent, on average, about 45 per cent of the total employment created. Table 23 shows the percentage by work organisation and by sector. The superseding of key-punch operators is confirmed by figures for the United States, where this occupation in data processing represented 29 per cent of total data processing personnel in 1974 and will go down to 18 per cent by 1985. The decrease in absolute numbers is from 249,000 to 200,000—that is to say, 19.7 per cent.[27]

Technological change is also having effects at the service, maintenance and manufacturing levels, and most of the new equipment does not need a specially protected environment. In the long term, paper will be used much less than today and thus many of the peripherals in current use will also be superseded, or at any rate the demand for them will be restricted. When all these elements are taken into account, it appears clear that the employment gain of the past will vanish and that the net result will be a loss of jobs coupled with a loss of potential job creation.

The analysis of employment in data processing is useful despite the fact that such employment represents a very small section of the labour force. Most other types of applications lead to a net loss of jobs, as in the case of office automation and after product and process changes in manufacturing. In general, other departments of an enterprise or institution have to adjust to data processing

Table 23.   Percentage of key-punch and verifier operators in total automatic data processing personnel in India

Number of shifts	Manufacturing	Educational or research and development
1	49	49
2	50	46
3	41	35

Source: Based on Prasad and Verma, *Impact of computers on employment*, op. cit., p. 62.

change, which is a powerful means of improving efficiency. In most cases this adjustment entails job losses.

Imports of computers and related equipment make further strains on scarce foreign exchange resources. As shown in table 24, imports of data processing equipment by developing countries have increased considerably. This is an important consideration not only in terms of job creation but also in framing a policy on computer use. In most developing countries both industry and services are sharply divided between a modern ("formal") sector and an "informal" sector consisting of a multitude of small businesses with extremely low productivity. This formal-informal division makes the employment problem qualitatively different from that of the developed countries.[28] A reduction or loss of potential employment in the formal sector increases pressure on the informal one and leads to further underemployment and unemployment which maintain low productivity and inhibit growth. In this classical circular situation the introduction of labour-saving technology has a chain effect and cannot be analysed only at the level of the individual enterprise using it. This aspect is particularly important when the equipment is imported, so that its manufacture has not contributed to the development of the domestic economy. Evidence accumulated in the industrialised market economy countries shows that information technology is encouraging business concentration. The reasons for this range from product changes which eliminate much subcontracting to the fact that it allows centralisation. Sometimes concentration is seen as bad in itself, although it can increase efficiency, in particular by pooling scarce resources and eliminating duplication. Nevertheless, negative effects can occur in developing countries when a practical policy of mixed technologies is not followed. In the context of the dichotomy between the formal and informal sectors, business concentration has a serious effect by forcing a decrease in the number of small businesses. Furthermore, the alteration of products produces a change in subcontracting arrangements for repair and maintenance which are an important source of employment for the informal sector.

By affecting employment and access to the means of production these changes have a considerable effect on income distribution. The unevenness of the distribution of income in turn inhibits development prospects by restricting the potential market and maintaining unacceptable levels of poverty, with profound economic, social and political consequences. The World Bank has stated that about 40 per cent of the population of developing countries are still

Table 24. OECD imports (cif) and exports (fob) of electronic products from and to developing countries, 1972-76
(Millions of US dollars)

Standard International Trade Classification items	1972		1976	
	Imports	Exports	Imports	Exports
7293	346	393	1 357	792
724–	208	198	245[1]	441
7242	295	202	831	669
7249	183	1 306	742	3 775
7142	12	188	245	669[1]
7143	19	126	37	265
8911	40	178	275	298[1]
Total	1 103	2 591	3 732	6 929

[1] 1975.
Key to items:
7293. Thermionic valves and tubes, transistors, etc. [photocells, integrated circuits]
724–. Telecommunications apparatus
7242. Radio broadcast receivers
7249. Telecommunications equipment n.e.s.
7142. Calculating and accounting machines etc. [including electronic computers]
7143. Statistical machines—cards or tapes
8911. Phonographs, tape and other sound recorders, etc.
Source: Organisation for Economic Co-operation and Development, Department of Economics and Statistics: *Statistics of foreign trade*, Series C: *Trade by commodities: Market summaries: General-imports-exports*, 1972, 1975 and 1976.

living in absolute poverty, and that the poorer members of the population in such countries are unlikely to share equitably in economic growth, mainly because they have less access to the productive assets needed to generate income.[29] In such circumstances the introduction of advanced technology operates differently inasmuch as it reinforces structural constraints on development. This calls for a careful evaluation of the introduction of any type of advanced technology, but particularly of information technology because its effects are felt primarily in the service sector which has been the job creator *par excellence*. On the other hand, proper evaluation and careful introduction of information technology can help to solve some of the problems mentioned above, provided that there is a co-ordinated effort by governments (and also international agreements in some respects) to set guidelines for the introduction and use of the technology. Up to now, developing countries have been slow to react to current change. Since computers have so far affected only a small number of workers, it is assumed, wrongly, that the number affected will remain small. Increasing international competition will force industry to rationalise further. When this happens and is combined with a loss of job creation potential in the service sector, employment prospects will be worse.

Notes

[1] Figures derived from Santosh Mukherjee: *Restructuring of industrial economies and trade with developing countries* (Geneva, ILO, 1978), p. 87. Mr. R. Kaplinsky of the Institute of Development

Studies, University of Sussex, has tentatively suggested that if an uneven diffusion of micro-electronic technology develops in the footwear industry, the effects will be similar to those detected in textiles, garments and electronics, i.e. the erosion of the competitiveness of developing countries.

[2] ILO: *Conditions of work in the textile industry, including problems related to organisation of work*, Report III, Textiles Committee, Tenth Session, Geneva, 1978, p. 43.

[3] See on this point European Communities, European Parliament, Working Documents, 1977-1978, Document 438/77, 13 Dec. 1977: *Report drawn up on behalf of the Committee on Economic and Monetary Affairs on the crisis in the textile industry* . . . (rapporteur: M. T. Normanton) and ILO: *Conditions of work in the textile industry*, op. cit., pp. 42, 65.

[4] Vincent Cable: *World textile trade and production, The Economist* Intelligence Unit Special Report No. 63 (London, 1979).

[5] This point has been continuously present in the debate about the textile industry in Europe and the United States; see Association of Scientific, Technical and Management Staffs discussion document: *Technological change and collective bargaining* (London, 1978; mimeographed), p. 20; also McLean and Rush, *The impact of microelectronics on the UK*, op. cit., pp. 21-28. For the United States see a survey conducted by *Business Week*, 14 May 1979, pp. 60-70.

[6] ILO: *Conditions of work in the textile industry*, op. cit., p. 43.

[7] Wilhelm Hardt: *Die Textilindustrie im Jahr 2000*, International Wool Conference, Basle, 8-12 June 1976 (mimeographed), as quoted in ILO: *Training requirements in the textile industry in the light of changes in the occupational structure*, Report II, Textiles Committee, Tenth Session, Geneva 1978, p. 9.

[8] Cable, *World textile trade and production*, op. cit., p. 37.

[9] UNCTAD, *Electronics in developing countries*, op. cit., p. V.

[10] Oldham, "The fabrication of microelectronic circuits", op. cit., p. 128.

[11] Heinz F. L. Roessle: "Prospects for the European semiconductor industry", in *Speakers' papers, Financial Times* conference, op. cit., p. 97.

[12] France, the Federal Republic of Germany, Ireland and the United Kingdom have designed special programmes to attract investment in micro-electronics. An important study in this field is from the Scottish Development Agency: *The electronics industry in Scotland: A proposed strategy*, by Booz, Allen and Hamilton, management consultants (1979).

[13] The main companies which have set up manufacturing facilities in the United States are Fujitsu, Hitachi, Matsushita, Nippon Electric Co., Sanyo, Sharp, TDK, Toshiba and Tri-Kenwood. By 1980, six Japanese television manufacturers will be producing 3 million sets a year in the United States, more than Japan exported in 1976, the peak year before the United States introduced protectionist measures.

[14] Electronics Industries Association of Korea (EIAK): *Official statistics* (Seoul, 1979).

[15] See for instance Government of India, Department of Electronics: *Annual report 1977-78* (New Delhi, 1978), and UNCTAD, *Electronics in developing countries*, op. cit.

[16] The evidence for this point is substantial. See Lamborghini, *The diffusion of microelectronics in industrial companies*, op. cit.; Werner Dostal: *Microelectronics and occupational skill* (1979; mimeographed); Hines and Searle, *Automatic unemployment*, op. cit., Ch. 2.

[17] *New Scientist*, 11 Jan. 1979, p. 79; see also Ch. 3 of the present work, note 29.

[18] Hans Peter Gassmann: "Data networks: New information infrastructure", in *OECD Observer*, Nov. 1978, pp. 11-16, and OECD, *The use of international data networks in Europe*, op. cit.

[19] J. M. Carroll: "The problem of transnational data flow", in OECD: *Policy issues in data protection and privacy*, op. cit., p. 203.

[20] Quoted by Gassmann, "New international policy implications of the rapid growth of transborder data flows", op. cit., p. 57.

[21] S. Sitaraman: "Export of computer software: A case study", in *Indian Management*, Aug. 1978, pp. 37-39.

[22] Dr. Koji Kobayashi, chairman of Nippon Electric Co., has stated that "the mental habits by which we have always associated electricity with copper wires will undergo a change". See K. Kobayashi: "The Japanese telephone industry in the year 2000", in ITU, *Telecommunications perspectives and economic implications*, op. cit., p. II.6.3. See also M. Hewish: "Aircraft designers follow the birds", in *New Scientist*, 4 Oct. 1979, p. 33, and "Lightweight composites are displacing metals", in *Business Week*, 30 July 1979, p. 36-D.

[23] Data supplied by the United States Bureau of Labor Statistics, as analysed in American Federation of Information Processing Societies (AFIPS): *Information processing in the United*

States, A quantitative summary, quoted by Pender M. McCarter: "Where is the industry going?", in *Datamation*, Feb. 1978, p. 104.

[24] Kamta Prasad and Pramod Verma: *Impact of computers on employment* (Delhi, Macmillan, 1976), p. 126.

[25] For the United Kingdom, Ministry of Labour, Manpower Research Unit: *Computers in offices*, Manpower Studies No. 4 (London, H M Stationery Office, 1965). For India, Government of India, Ministry of Labour and Rehabilitation: *Report of the Committee on Automation* (Chairman: Prof. V. M. Dandekar) (1972; mimeographed).

[26] ILO: *Automation in developing countries*, Round-table discussion on the manpower problems associated with the introduction of automation and advanced technology in developing countries Geneva, 1-3 July 1970), p. 215.

[27] McCarter, "Where is the industry going?", loc. cit.

[28] Numerous works have been published covering this point. Apart from the reports of comprehensive employment strategy missions undertaken by United Nations inter-agency teams organised by the International Labour Office and financed by the United Nations Development Programme—*Towards full employment: . . . Colombia* (1970), *Matching employment opportunities and expectations: . . . Ceylon* (1971), *Employment, incomes and equality: . . . Kenya* (1972), *Employment and income policies for Iran* (1973), *Sharing in development: . . . the Philippines* (1974), *Growth, employment and equity: . . . the Sudan* (1976)—all published by the ILO in Geneva, see also Henry Rempel and William J. House: *The Kenya employment problem: An analysis of the modern sector labour market* (Nairobi, Oxford University Press, 1978); and for Latin America, Oficina Internacional del Trabajo, Programa Regional del Empleo para América Latina y el Caribe (PREALC): *Políticas de empleo en América Latina*, and *Sector informal: Funcionamiento y políticas* (Santiago, Chile, 1974 and 1978 respectively).

[29] The World Bank (Washington): *World development report, 1978*, p. 7.

SUMMARY AND CONCLUSIONS

The material presented in this book covers two inter-related subjects: the general and abstract features of information technology, and its practical implications. From this material the following conclusions can be drawn.

The revolutionary nature of current changes finds its ultimate expression in the achievement of a comprehensive information handling system which uses a uniform type of signal with binary values: the electronic bit or basic unit of information. This achievement has been made possible through the evolution of semiconductor technology, particularly integrated circuits, in which the number of functions per chip has increased exponentially. That development has been followed by the invention of the microprocessor and microcomputer, which process information at high speed and with great accuracy, and constitute the basic components of information technology. Electronic components, computers and telecommunications can now be brought together under the new concept of telematics, which denotes the combination of information technology and telecommunications.

Being basically concerned with handling, processing, storing, and retrieving information, information technology is related in many ways to the functions of human intelligence. This fact marks a new and different progression over historical discoveries which led to the multiplication of human muscle power and to the mastery of new sources of energy as a substitute for it. Like the human brain, the microprocessor and microcomputer are universal information-processing devices that can be used to perform a wide variety of functions. No mechanical or intellectual activity can take place without some form of information exchange: this explains the pervasiveness of micro-electronic technology.

The need for information technology stems from two inter-related factors. First, information is accumulating at such a rate in all fields that in the highly industrialised countries an increasing proportion of the labour force is engaged in handling information. This trend is even becoming evident in some advanced developing countries. Secondly, there is a need to increase productivity in office and clerical work since economic growth can no longer be achieved by an increase in industrial and agricultural productivity alone; in the industrialised

103

market economy countries automation is necessary in all sectors of the economy to maintain international competitiveness.

Coupled with the nature of the technology are its technical characteristics and particularly its economy. The prices of components, computer power and telecommunications are falling at unprecedented speed, and this fall explains the speed of diffusion and pervasiveness of micro-electronics.

It is difficult to find a field in which the processing power of micro-electronics cannot be used. The applications affect—

manufacturing, by altering products and processes;

office work, by rationalising routine information-handling operations and by increasing the independence of office workers from former channels of communication;

services, by increasing automation, self-service and the substitution of goods for person-to-person services; and

information flows, by altering the information infrastructure, the speed of handling information and the development of national and international data networks.

The combination of these elements clearly affects the individual and socio-economic and political structures at the national and international levels.

By any standard, information technology—components, computers and telecommunications—is spreading very fast. Electronics is the fastest-growing industry in almost all industrialised countries, and out-performs the general growth rate of the economy. The main force behind the rapid diffusion of the technology is the need for the firms concerned to remain internationally competitive, and thus market penetration in any one country depends on the behaviour of competitors.

Nevertheless, a number of factors act as brakes. From an economic point of view the single most important check is the need to amortise old equipment and the over-all cost of new applications. The inadequacy of existing telecommunications facilities will also slow down applications. Traditional and deep-rooted managerial practices also act as a brake, coupled with legal restrictions in some fields. The basic attitude of trade unions has been one of "change by consent", and the reaction of the general public to further centralisation, threats to privacy and so on will regulate rather than slow down the application of the new information systems.

In developing countries the technology is diffusing rapidly but is mainly restricted to traditional applications. The main constraints on diffusion are: the low cost of labour, which renders the equipment less competitive and makes amortisation take longer; government policy on tariffs, applications, technology transfer and telecommunications; and the lack of skills, particularly for non-conventional applications and production processes.

The full development of information technology is possible only in a world market. This constraint is due to the economies of scale required to amortise the tremendous research and development expenses which a firm must incur and the enormous capital investment it must make in order to meet keen competition. For computers, components and telecommunications, all the main producers are multinational and the market is highly concentrated; companies incorporated in the United States are particularly strong. Data banks, software and machine services are also highly concentrated. The structure of the industry and its heavy dependence on a technological edge makes technology transfer or a more even distribution of manufacturing facilities unlikely. On the contrary the current tendency, due essentially to capital requirements, is towards further concentration.

The foregoing remarks show that electronics is becoming a convergence industry: its applications are tending to determine where innovations shall take place; it is becoming essential for an ever-growing number of industries; and it is conditioning skill requirements and manning levels. In fact electronics, in its varied expressions, will substantially condition industrial and service activities and the socio-political structure; it will affect both the ways in which people produce and the ways in which they relate to each other.

From a socio-economic point of view, the most important effect in the short run will be on employment. This must be evaluated in terms of current circumstances, that is to say lower growth prospects, a transition to new forms of energy, increase in the labour force, recession, inflation and already high unemployment. The first sector to be affected in a quantitative or qualitative form will be clerical work in offices and services. This inevitably means that women workers will be deeply affected. Industry will be relatively less affected because of its traditionally higher productivity levels, but even there workforce reductions will be substantial in absolute terms as a result of the alteration of products and processes. Although the use of the technology generally leads to savings in capital per unit of output, it sharply increases the ratio of capital to labour. Two simultaneous employment effects of the introduction of the new technology are labour displacement and a loss of job creation potential; they lead to an over-all decrease in the labour requirements of the economy. Electronic-based products and industries are basically labour-saving. Even a sharp increase in the output of new products will not necessarily alter the long-term trend, as has been shown by the performance of the electronics industry itself: in most cases there is no need to develop new distribution channels and repair and maintenance facilities.

A quantitative assessment of the employment effects meets with tremendous methodological problems owing to the lack of refined quantitative methods and the difficulties in isolating technological from other effects. However, from the imperfect figures which do exist on job creation and job displacement, and on the effects that have already occurred in many industries and services, it is clear that a substantial loss of jobs is taking place, and will accelerate in the early 1980s. The scope afforded by international and national competition, together with the new needs generated by the new technology, will offset job losses to some extent, but not in a substantial manner. These developments are consistent with the historical fall in employment in agriculture and the more recent falls in industry and now in services.

It would seem that a transition is taking place from a society with unemployment to one that no longer needs its full potential labour force to produce the necessary goods and services under current conditions of work. It is doubtful whether measures such as early retirement, shorter working hours, and the creation and development of small businesses and new products and services will have much effect on job creation. Nevertheless, the need to meet the educational, cultural and social needs inherent in such a transition, plus cultural resistance, could lead to job creation in some fields. A transition of this nature will not be free from turmoil while the population tries to adjust to new life styles.

There will be changes in working conditions due to new routines, isolation and health hazards—the latter due to the extensive use of terminals with visual display screens. In addition, important changes will take place in the structure of enterprises and the skill levels of large sections of the labour force. The traditional distinctions between blue-collar and white-collar workers will diminish. In the long run this change will produce modifications in trade union structure and in the patterns of social mobility. Most important will be the effects on society and income distribution when much of the population is not employed for long periods of time or at all.

Appropriate forms of social control will be required to ensure that there are no violations of privacy and other human rights and that knowledge will not be concentrated in a small technocratic élite, thus polarising the political system and rarefying the political debate. On the other hand, responsible use of new systems and devices could lead to democratisation and political and administrative as well as physical decentralisation.

Owing to the development of national and international data networks, dependence among countries sharing computer resources is increasing. Transborder data flows are intensifying the concentration of information-intensive activities in some developed countries: while data providing the basis for decision-making will flow towards the developed countries, information flowing towards the developing countries will embody decisions already taken. "Electronic consultancy" and transfer of data to be processed abroad involve a loss of local income and jobs, since in economic terms they amount to the export of clerical and scientific work, a sort of "electronic brain drain". There are also important implications in relation to transfer of technology: hitherto, by and large, this has been a question of equipment or "hardware"; but the world is increasingly faced with the transnational flow of an intangible good, information, which is of tremendous economic value, although it is difficult to identify and its movements are not easy to trace. At the same time, information reflects ethical judgements or values, and embodies concepts and interests which may not correspond to the needs and cultural identity of various nations and other groups.

Information technology will affect the international division of labour in the following ways:

1. The increase in automation lessens the importance of direct labour costs in total production costs, thus making the manufacture of formerly labour-intensive goods economically feasible in developed economies. This effect is already apparent with regard to goods such as textiles, garments and electronic products, which the developing countries used to export in large quantities to the economically more developed part of the world. In these cases the competitive advantage of less developed countries is being eroded through automation, and some key industries are returning to the developed countries. In addition, some

industries which were potential candidates for transfer to developing countries will now be able to survive in the developed countries with the help of automation.

2. The characteristics of the information processing technology reinforce some of the main advantages of the industrialised market economy countries, namely advanced management techniques, extensive co-ordination, efficiency and systematic marketing.

3. Traditional mass production industries (e.g. textiles and garments) are increasingly launching out into research and development and becoming capital-intensive, thus entering the high-technology category, in which design and quality control are essential to the acceptability of products.

4. The technology permits further industrial and service concentration and vertical (forward) integration, which in turn implies streamlining in certain sectors, the accumulation of resources and an increase in the resources and marketing drive of multinational companies. However, the loss of production flexibility and innovative capacity in industrial giants creates considerable opportunities for newcomers. These opportunities may be seized and exploited, at least for a while, by small technology-based businesses. However, for the reasons already indicated, they will establish themselves not in the developing countries but in those that are already developed.

5. The shift from labour-intensive to capital-intensive activities, which is implied by the use of micro-electronic technology, which affect capital requirements and capital formation.

The foregoing trends suggest that the possibility of transferring industries from the advanced industrialised countries to the developing countries is far more remote than had previously been assumed. With few exceptions, it is unlikely that developing countries will be able to combine high technology with cheap labour. There are numerous factors involved: among the most intractable are the capital and skill requirements needed to successfully exploit micro-electronic technology at the product or process level. Current changes are software-based; the elaboration and use of software requires sophisticated skills and experience accumulated in solving previous programming problems. At the same time, the electronics industry is coming to rely less on mastering the design and manufacture of individual parts than on system engineering, which is based on a convergence of skills that are much more sophisticated and difficult to acquire. Yet the proper use of the potential of available technologies could help solve some of the most pressing and very diverse problems of the developing countries by optimising the allocation of scarce resources and by facilitating improved education and training, rapid and economic access to vast pools of knowledge, and access to cheaper, more accurate and reliable machine tools and equipment.

Current changes are also tending to produce others in such fields as materials (the substitution of fibre optics for copper), sensory devices and opto-electronics. It is not yet known how materials substitution will affect raw material producers, and this is a matter which deserves urgent consideration. Breakthroughs in sensory devices and opto-electronics, and further advances in

107

hydraulics and pneumatics, will multiply applications and open the way to further automation.

Most developing countries have followed the development rationale of the already highly industrialised countries, and tend to overlook the restrictions built into the existing structure of the world economy. In developed countries also, in the majority of cases, due weight has not been given, in the debate about the "post-industrial society", to the interdependence of developed and developing countries, and only a sort of "paragraph hommage" has been paid to possible trends in developing countries. Owing to the dynamism of technological change, which may accelerate in crucial areas in the future, comparative advantages such as low labour costs are no longer the most important asset for many developing countries. This situation will further encourage the need to look for other development paths. More often than not, this need is recognised but little is done to pursue its implications, particularly in relation to an alternative concept of science and technology. The slow reaction of developing countries on issues such as transborder data flows and their effects provides an example of a case in which developing countries will in the end have almost no choice but to follow the systems developed in advanced countries. This also involves the absorption of information that is not value-free and corresponds to a particular concept and method of assessing and pursuing progress.

The failure to recognise these problems is rooted in a deeper theoretical issue in economic studies, namely that technological change has been regarded as an exogenous factor in economic analysis (either completely exogenous or incorporated in capital and/or labour). It is indeed difficult to incorporate technology as an endogenous variable because of the intrinsic complexities of the changes themselves and the need for technological forecasting. Despite these difficulties, forecasting is becoming essential at every level, from investment planning to the establishment of different aspects of economic or industrial policy. Planning techniques, which are still in their infancy, are nevertheless more developed in advanced countries, where management and governments now feel bound to act on possible trends (technological, market, etc.). A similar policy is seldom followed in developing countries, which are "information-poor" by comparison with the "information-rich" highly industrialised countries.

In the 1980s, now that the blossoming of information technology is expected, the breadth of the resulting problems will be international. This universal scope is due to the interdependence of the world economy, and also to the fact that the technology cannot be profitably applied without a world market. In view of this situation, it is in the best interests of developing and developed countries alike to confront the issues as a matter of urgency. The erosion of the competitiveness of the developing countries may lead to a fragmentation of world trade, to more insoluble social and economic problems and thus to more difficult relations between the Northern and the Southern hemisphere. This difficulty will be increased by a less favourable attitude in developed countries towards international co-operation because of their own internal problems and competition among themselves. The lack of progress in the North-South dialogue shows that there are no easy answers to the situation, but neither is it helpful to dismiss further complications in the hope that they might simply go away.

One way of easing some of the difficulties would clearly be the development of a more comprehensive body of knowledge in relation to science and technology in general and its effect on the fabric of society and the international and national division of labour in particular. In this respect, information technology is a case study in a much broader field that calls for further attention. The advent of information technology is full of opportunities and also dangers; the direction it finally takes will depend largely on action at the national and international level and joint assessment among users, producers, governments, unions and academics. What is clear at this stage is that, given current trends, there is no time to be lost.

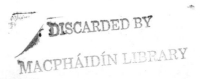